eTELEMED 2021

The Thirteenth International Conference on eHealth, Telemedicine, and Social Medicine

July 18 – 22, 2021

Nice, France

eTELEMED 2021 Editors

Prima Oky Dicky Ardiansyah, Iwate Prefectural University, Japan
Arian Rajh, Agency for Medicinal Products and Medical Devices of the Republic of Croatia, Croatia

Printed from e-media with permission by:

Curran Associates, Inc.
57 Morehouse Lane
Red Hook, NY 12571

CURRAN ASSOCIATES INC. proceedings .com

Some format issues inherent in the e-media version may also appear in this print version.

International Academy, Research, and Industry Association (IARIA)
412 Derby Way
Wilmington, DE 19810

Phone: (408) 893-6407
Fax: (408) 527-6351

petre@iaria.org

Additional copies of this publication are available from:

Curran Associates, Inc.
57 Morehouse Lane
Red Hook, NY 12571 USA
Phone: 845-758-0400
Fax: 845-758-2633
Email: curran@proceedings.com
Web: www.proceedings.com

Table of Contents

eTELEMED 2021

Forward

The Thirteenth International Conference on eHealth, Telemedicine, and Social Medicine (eTELEMED 2021), held in Nice, France, July 18 - 22, 2021, considered advances in techniques, services, and applications dedicated to a global approach of eHealth.

Development of wireless homecare, of special types of communications with patient data, of videoconferencing and telepresence, and the progress in image processing and date protection increased the eHealth applications and services, and extended Internet-based patient coverage areas. Social and economic aspects as well as the integration of classical systems with the telemedicine systems are still challenging issues.

eTELEMED 2021 provided a forum where researchers were able to present recent research results and new research problems and directions related to them. The topics covered aspects from classical medicine and eHealth integration, systems and communication, devices, and applications.

We take this opportunity to thank all the members of the eTELEMED 2021 Technical Program Committee as well as the numerous reviewers. The creation of such a broad and high-quality conference program would not have been possible without their involvement. We also kindly thank all the authors who dedicated much of their time and efforts to contribute to the eTELEMED 2021. We truly believe that, thanks to all these efforts, the final conference program consists of top quality contributions.

This event could also not have been a reality without the support of many individuals, organizations, and sponsors. We are grateful to the members of the eTELEMED 2021 organizing committee for their help in handling the logistics and for their work to make this professional meeting a success.

We hope that eTELEMED 2021 was a successful international forum for the exchange of ideas and results between academia and industry and for the promotion of progress in eHealth and Telemedicine research.

eTELEMED 2021 Chairs

eTELEMED Steering Committee
Yoshitoshi Murata, Iwate Prefectural University, Japan
Arian Rajh, Agency for Medicinal Products and Medical Devices of the Republic of Croatia, Croatia

eTELEMED 2021

COMMITTEE

eTELEMED Steering Committee

Yoshitoshi Murata, Iwate Prefectural University, Japan
Arian Rajh, Agency for Medicinal Products and Medical Devices of the Republic of Croatia, Croatia

eTELEMED 2021 Advisory Committee

Lisette Van Gemert-Pijnen, University of Twente - Enschede, the Netherlands
Marike Hettinga, Windesheim University of Applied Sciences, the Netherlands
Åsa Smedberg, Stockholm University, Sweden
Conceição Granja, Norwegian Centre for e-health Research, University Hospital of North Norway, Norway

eTELEMED 2021 Publicity Chair

Alvaro Liebana, Universitat Politecnica de Valencia, Spain
Mar Parra, Universitat Politecnica de Valencia, Spain

eTELEMED 2021 Technical Program Committee

Don Adjeroh, West Virginia University, USA
Giovanni Albani, Istituto Auxologico Italiano - IRCCS, Italy
Prima Oky Dicky Ardiansyah, Iwate Prefectural University, Japan
Theodoros N. Arvanitis, Institute of Digital Healthcare | University of Warwick, UK
Tunç Aşuroğlu, Başkent University, Ankara, Turkey
Rafael Avila, Hospital Universitario Privado de Cordoba, Argentina
Mansoor Baig, King Faisal Specialist Hospital & Research Center, Riyadh, Saudi Arabia / ICIMTH, Greece
Panagiotis D. Bamidis, School of Medicine - Aristotle University of Thessaloniki, Greece
Oresti Baños, University of Granada, Spain
Ivana Bartoletti, Gemserv, UK
Fernández Batalla Marta, University of Alcala, Spain
Christian-Alexander Behrendt, GermanVasc Research Group/ University Medical Center Hamburg-Eppendorf, Germany
Elisabetta Benevento, University of Pisa, Italy
Arriel Benis, HIT- Holon Institute of Technology, Israel
Vilmos Bilicki, University of Szeged, Hungary
Lucia Billeci, National Research Council of Italy | Institute of Clinical Physiology, Pisa, Italy
Enrico Gianluca Caiani, Politecnico di Milano, Italy
Manuel Campos Martínez, University of Murcia, Spain
Nicola Carbonaro, University of Pisa, Italy
Ayan Chatterjee, The University of Agder, Grimstad, Norway
Darwyn Chern, Copa Health, Phoenix, USA
Bhargava Chinni, University of Rochester, USA
Mario Ciampi, National Research Council of Italy | Institute for High Performance Computing and

Newtorking, Italy
James J. Cimino, Informatics Institute - University of Alabama at Birmingham, USA
Javier Civit, Cober SL, Spain / Gnomon Informatics, Greece
Daniel Condor Camara, Cayetano Heredia University, Peru
Massimo Conti, Università Politecnica delle Marche, Ancona, Italy
Sandra Costanzo, University of Calabria, Italy
Paul M. Cunningham, IST-Africa Institute, Ireland
Jacques Demongeot, Université Grenoble Alpes, France
Pierpaolo Di Bitonto, Grifo multimedia S.r.l., Italy
Gayo Diallo, Univ. Bordeaux/ISPED, France
Linying (Lin) Dong, Ryerson University, Canada
Audrey DunnGalvin, University College Cork, Ireland
Duarte Duque, 2Ai | Polytechnic Institute of Cávado and Ave Barcelos, Portugal
Claudio Eccher, FBK Fondazione Bruno Kessler, Italy
Dina El Demellawy, CHEO Research Institute | University of Ottawa, Canada
Christo El Morr, York University, Canada
Radwa El Shawi, University of Tartu, Estonia
Bruno Fionda, Fondazione Policlinico Universitario "A. Gemelli" IRCCS, Italy
Sebastian Fudickar, Universität Oldenburg, Germany
Niels F. Garmann-Johnsen, University of Agder, Norway
Anthony Gelibert, Carbon Bee, France
Wojciech Glinkowski, Polish Telemedicine Society / Center of Excellence "TeleOrto", Poland
Kuang Gong, Massachusetts General Hospital / Harvard Medical School, USA
Manuel González-Hidalgo, University of the Balearic Islands / Balearic Islands Health Research Institute (IdISBa), Spain
Adela Grando, Arizona State University | College of Health Solutions, USA
Conceição Granja, Norwegian Centre for e-health Research, University Hospital of North Norway, Norway
David Greenhalgh, University of Strathclyde, UK
Teresa Guarda, Universidad Estatal Peninsula Santa Elena - UPSE / Universidad de las Fuerzas Armadas – ESPE / ALGORITMI Research Centre | ESPE | UPSE, Ecuador
Katarina Gvozdanovic, Agency for Medicinal Products and Medical Devices, Zagreb, Croatia
Oliver Heinze, University Hospital Heidelberg, Germany
Pilar Herrero, Universidad Politécnica de Madrid, Spain
Marike Hettinga, Windesheim University of Applied Sciences, the Netherlands
Delowar Hossain, BRAC University | United International University | IDCL Fiance Limited, Bangladesh
Amin Hossein, Université libre de Bruxelles, Belgium
Ying-Feng Hsu, Osaka University, Japan
Yan Hu, Blekinge Institute of Technology, Sweden
Ming Huang, Mayo Clinic, USA
Femi Isiaq, Southampton Solent University, UK
Ashad Kabir, Charles Sturt University, Australia
Martijn Kiers, University of Applied Science FH JOANNEUM, Austria
Toralf Kirsten, University of Applied Sciences Mittweida, Germany
Evdokimos Konstantinidis, Aristotle University of Thessaloniki, Greece / Nively, Nice, France
Frank Kramer, Faculty of Medicine/University of Augsburg, Germany
Vinay Kumar, Thapar University, Patiala,India
Siru Liu, University of Utah, USA

Copyright Information

3D Human Pose Estimation of a Partial Body from a Single Image and Its Application in the Detection of Deterioration in Sitting Postures

Oky Dicky Ardiansyah Prima and Kazuki Hosogoe
Graduate School of Software and Information Science, Iwate Prefectural University
152-52 Sugo Takizawa, Japan
email: prima@iwate-pu.ac.jp, g231r026@s.iwate-pu.ac.jp

***Abstract*—Three-dimensional (3D) human pose estimation has been used in a wide range of fields, including motion analysis in sports and rehabilitation, modeling in Computer Graphics (CG) production for movies and games, and input interfaces. Recently, 3D human pose can be estimated with high accuracy only from a single image using a neural network model. However, depending on the camera's position and shooting angle, some joints may be occluded, thus reducing the accuracy of the overall joint estimation. In this study, we experimentally constructed a neural network model for 3D human pose estimation based on a single image and evaluated the difference in accuracy of the pose estimated by the model constructed for the partial joints of the body and the whole-body joints. The dataset used for the experiment was Human 3.6M and a human pose dataset created from an RGB-D camera for this study. The results confirmed that the model built based on the upper-body joints of the body had higher accuracy than that for the whole-body joints at estimating the posture of the upper body. Finally, we demonstrated that 3D human pose can be used to detect the deterioration in sitting postures, which can suggest that the technology is effective in improving various postures in daily life.**

Keywords-pose-estimation;3D human pose; body joint; computer vision.

I. INTRODUCTION

Three-Dimensional (3D) human pose estimation is the problem of locating the position of human joints in an image or video. Estimating human pose is an important problem that has received a lot of attention in the field of computer vision over the past few decades. Recently, improvements in 3D human pose estimation based on neural networks have led to higher accuracy and higher frame rates, as well as lower prices and simpler use of measurement equipment for human pose estimation [1][2]. As a result, 3D human pose estimation is being used in a wider range of fields, and the scope of its application is expanding not only to large-scale applications by companies and research institutions but also to personal use and other small-scale applications. A typical example of the use of human pose estimation technology at the individual level is video production using virtual characters whose movements are synchronized in real time. The use of such virtual characters with synchronized movements has been rapidly expanding in the entertainment field.

3D human pose information can be used for motion analysis to evaluate the effectiveness of rehabilitation. Using this information, it becomes convenient to measure the range of motion of the patient's joints and to perform self-rehabilitation at home. Experiments conducted by Prima et al. (2019) indicated that 3D human pose estimation from a single image is more advantageous for estimating semi-occluded joint locations than those estimated by depth cameras [3].

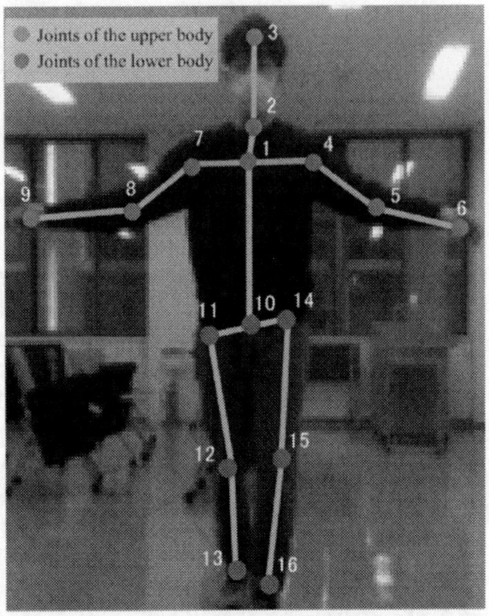

Figure 1. Sixteen joints extracted from the dataset for this study.

Existing studies on 3D human pose estimation have mainly focused on developing and verifying methods assuming that the whole-body joints can be captured. However, depending on the capturing environment and angle, a part of the body may be occluded and thus only some parts of the body can be captured. For example, when a person is sitting at a desk that is fully covered and photographed from the front, the lower half of the body is hidden by the desk and only the upper half of the body is captured. In such a situation, if 3D human pose estimation is performed for all joints of the body, the missing information of some joints may affect the detection of other joints and reduce the estimation accuracy.

This study aims to compare the accuracy of human 3D human pose estimation based on a single image using a model consisting of the whole-body joints and a model consisting of partial joints without occlusion, and to verify the effectiveness of the model corresponding to partial joints when a part of the body is occluded. As an application, a technique for detecting deterioration in sitting postures using 3D human pose estimation is introduced.

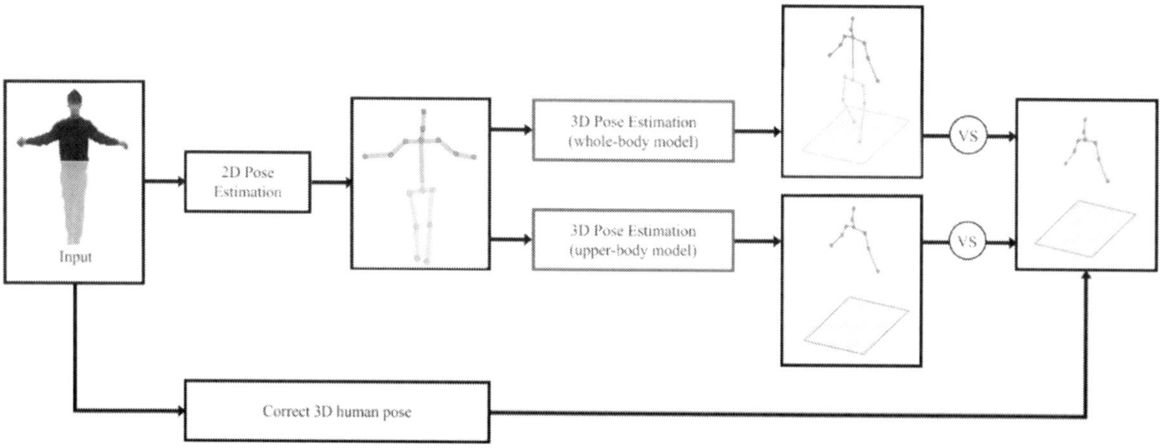

Figure 2. The evaluation flow of the 3D human pose acquired with the whole-body model and the upper-body model, respectively.

This paper is organized as follows. In Section II, we describe the 3D human pose models built for this study. Section III describes the method used to evaluate the accuracy of the partial 3D human pose. In Section IV, we present our results evaluated using the public dataset and the dataset created by an RGB-D camera. Section V introduces a technique for detecting deterioration in sitting postures using 3D human pose estimation. Finally, Section VI summarizes the results of this study and discusses future perspectives.

II. Building 3D Human Pose Models

In this study, we used the Human3.6M [4] to build 3D human pose models. Human3.6M is a 3D human pose dataset consisting of 3.6 million human poses and corresponding images, recorded from the performances of five female and six male subjects. This dataset consists of 2D joint positions, 3D joint positions, RGB images, time-of-flight (depth) data, and 3D body scan data from four different viewpoints measured by a high-speed motion capture system. Sixteen joints (Figure 1), were finally extracted from the dataset, and used in the evaluation of 3D human pose in this study. There are nine joints in the upper body and seven in the lower body.

We chose the 3D baseline method [1] to build our 3D human pose models because it can achieve low error rates on 3D human pose estimation using a relatively simple deep feed-forward neural network. The network expands the dimension of the input data to 1024 using the initial weights by He et al.'s (2015) [5], and then performs a series of batch normalization, Rectified Linear Unit (RELU), dropout rate of 0.5, and a residual connection. Following [1], subjects 1, 5, 6, 7 and 8 were used for training, and subjects 9 and 11 for evaluation. Two models, i.e., a 3D human pose model trained using joints of the whole body, and a model trained using only joints of the upper body were built for this study. Hereinafter, the former is referred to as the "whole-body model" and the latter as the "upper-body model."

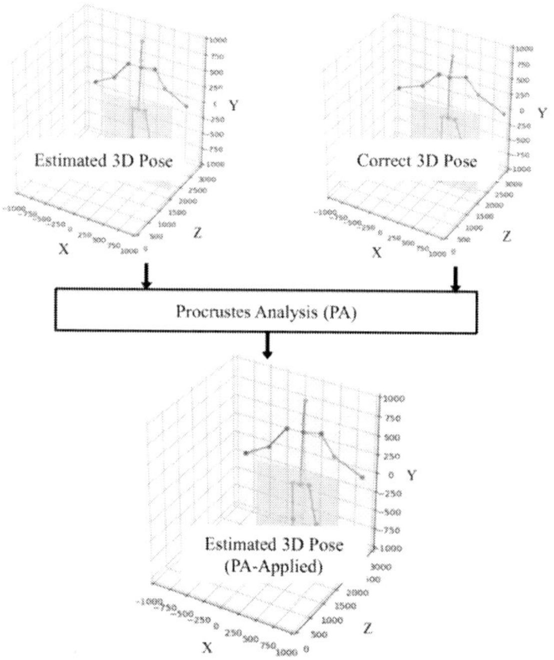

Figure 3. Procrustes analysis applied to the estimated 3D pose.

III. Evaluation of Partial 3D Human Poses

Our evaluation process goes as follows (Figure 2). After extracting the two-dimensional (2D) human pose from the image, we obtain the 3D human pose estimated by each model. For the full-body model, the 3D joints of the whole body are acquired, whereas for the upper-body model, the 3D joints of the upper body are acquired.

We then compare the obtained 3D human poses with the correct 3D human poses of the upper body and calculate the

TABLE I. ERRORS IN UPPER BODY POSES IN THE HUMAN3.6M ESTIMATED BY THE WHOLE-BODY AND THE UPPER-BODY MODELS.

No.	Joints	Model [mm]	
		Whole-Body	**Upper-Body**
1	Neck	162.79	51.24
2	Nose	38.39	26.71
3	Head	80.75	45.65
4	Left Shoulder	81.05	37.39
5	Left Wrist	91.10	45.13
6	Left Elbow	109.46	69.94
7	Right Shoulder	72.66	35.43
8	Right Wrist	89.63	46.64
9	Right Elbow	105.65	69.50
Mean (*M*)		92.387	47.514
Standard deviation (*SD*)		33.5448	14.5406

TABLE II. ERRORS BETWEEN THE 3D POSES OF THE UPPER BODY ESTIMATED BY THE WHOLE-BODY AND THE UPPER-BODY MODELS AND BY THE RGB-D CAMERA.

No.	Joints	Model [mm]	
		Whole-Body	**Upper-Body**
1	Neck	114.70	74.87
2	Nose	181.63	145.16
3	Head	96.70	**122.22**
4	Left Shoulder	103.78	91.44
5	Left Wrist	102.63	92.35
6	Left Elbow	148.85	107.67
7	Right Shoulder	107.51	**113.34**
8	Right Wrist	102.86	**108.82**
9	Right Elbow	158.45	111.21
Mean (*M*)		124.123	107.453
Standard deviation (*SD*)		30.7033	20.1145

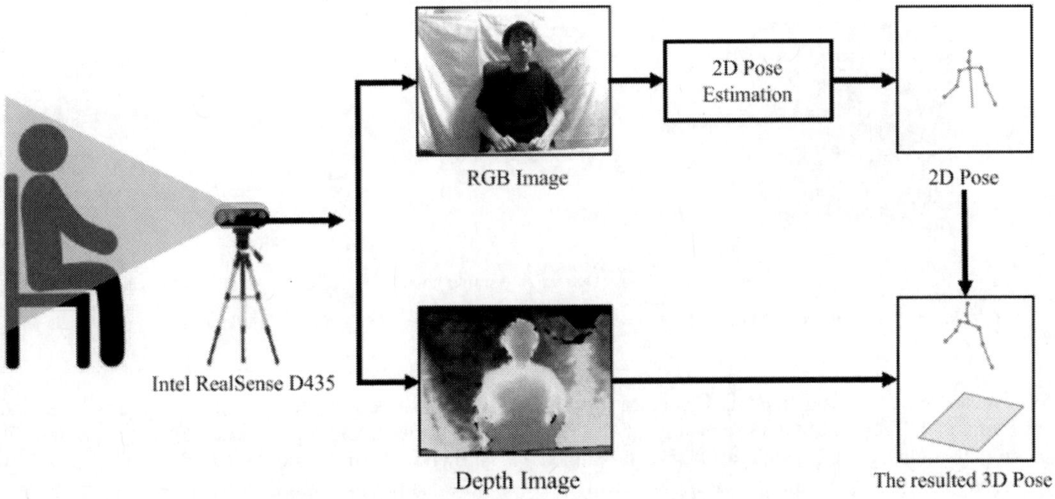

Figure 4. 3D human pose data generation using an RGB-D camera (Intel RealSense D435).

accuracy of the pose estimation by each model. Finally, we evaluate the difference in the accuracy of 3D human pose estimation between the model for the whole body and the model for the upper body in the situation where only the upper body is captured.

To find differences in poses, Procrustes Analysis (PA), a shape-preserving Euclidean transform, is applied to the estimated 3D human pose with reference to the correct one (Figure 3). This analysis eliminates the variation in movement, rotation, and scaling between the pose data while preserving the shape.

IV. RESULTS

To evaluate the difference in accuracy between the whole-body and the upper-body models, we randomly selected 548,800 human poses from the Human3.6M, which were not used for training and validation of these models. Table I shows the estimation errors of the upper body poses by each model. The accuracy of the estimation by the upper-body model is high for all joints. Overall, the accuracy of pose estimation significantly improved by about 45mm when the upper-body model was used to estimate the pose of the upper body (M = 92.387, SD = 33.5448, $t(8) = 4.99$, $p < .001$).

In order to confirm the results of estimating the 3D human pose in a real world, we further evaluated these models with

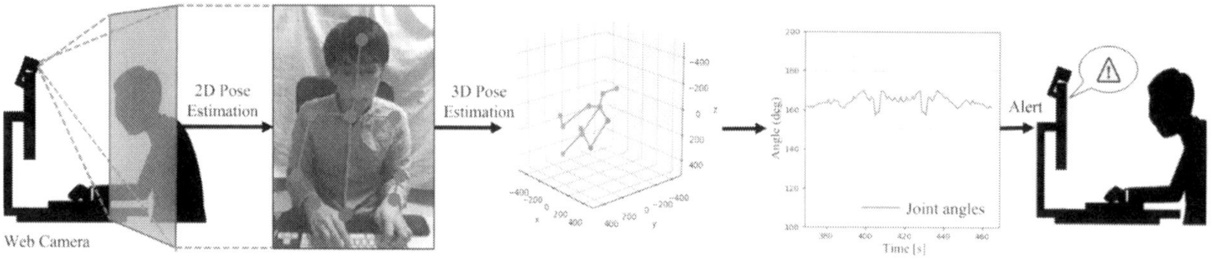

Figure 5. The detection of deterioration in sitting postures based on 3D human pose estimation.

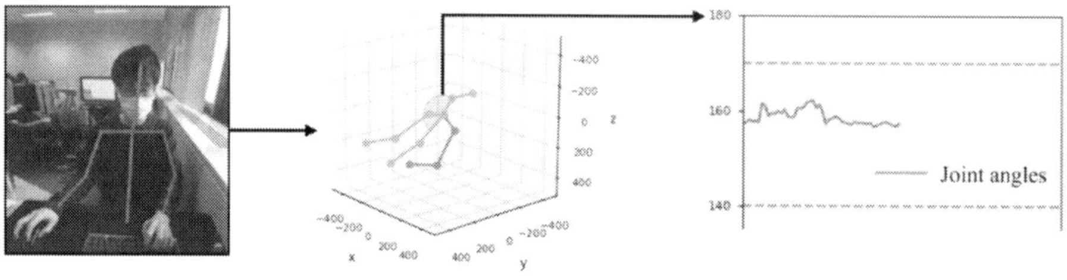

(a) Good sitting postures.

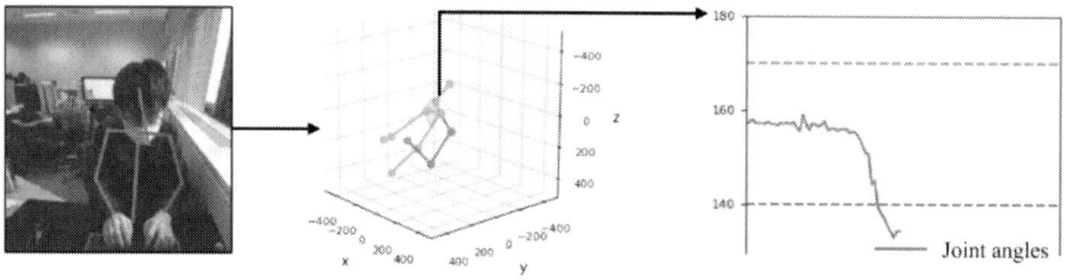

(b) Deteriorated sitting postures.

Figure 6. The difference in the angle formed by the nose, neck, and pelvis in good and deteriorated sitting postures.

the results of measurements using an RGB-D camera. For this study, we generated 7,350 3D human pose data from a subject using the Intel RealSense Depth Camera D435. The subject was seated about 1m from the camera and moved his hands and body in various ways as long as his body did not move out of frame from the camera.

To generate 3D human pose data from the D435, we first apply geometric corrections to the RGB and depth images from the camera to match the coverage of both images (Figure 4). For RGB images, the Stacked Hourglass Network [6] is used to estimate the 2D human pose. By superimposing the 2D human pose information with the depth image, the depth data of each joint is extracted to generate 3D human pose data. However, since the depth data is affected by noise which makes it difficult to obtain the accurate depth of the joint, we filtered this data using a five-frame moving average in time series. The camera's resolution was set to 640x480 pixels and the frame rate to 30fps.

Table II shows the estimation errors of the upper body posed by the whole-body and the upper-body models, calculated from the data acquired with the D435. Although the estimation errors in the head, right shoulder, and right wrist are higher than those estimated by the whole-body model, as shown in the bold numbers, the overall accuracy of 3D human pose estimation significantly improved by about 17mm when the upper-body model was used to estimate the pose of the upper body ($M = 124.123$, $SD = 30.7033$, $t(8) = 1.94$, $p < .05$).

V. Detection of Deterioration in Sitting Postures

During desk work, people tend to unconsciously adopt a posture that places a strong load on some part of their body. Slouching, a posture where the back is arched and the head is forward, is a typical poor posture when sitting. Slouching causes stiff shoulders and back pain due to the load placed on the neck, shoulders and back to support the weight of the head moving forward. Although the sitter can always be aware of and strive to improve their posture when sitting, it would be difficult for them to maintain such awareness during long hours of desk work.

In this study, we developed a method for detecting deteriorated sitting postures from images acquired from a

Web camera installed above a PC monitor using 3D human pose estimation. The advantage of a webcam is that it is widely used and installed as a standard feature of many laptops, allowing for easy access to our purpose. Furthermore, since there is no physical contact between the sensor device and the user, it is comfortable and does not interfere with movement (Figure 5). As joint angles can be calculated from 3D human pose, the characteristics of those angles in sitting postures can be analyzed.

To assess sitting postures, we focused on the change in the angle formed by the nose, neck, and pelvis. As shown in Figure 6, two different sitting postures change the angles formed by the nose, neck and pelvis. These angles change significantly as the head moves forward. The dotted line shows the empirical threshold for the tolerance of good sitting posture. As the sitting posture deteriorated, the observed angle values would no longer fall within this acceptable range.

VI. Conclusion

In this study, we experimentally constructed a 3D human pose estimation model specialized for a part of the body using a single camera. As a result, we confirmed that the accuracy of 3D human pose estimation by the model specialized for given parts of the body was higher than that by the model for the whole body.

The proposed system to detect posture deterioration during sitting using the 3D human pose model specialized for the upper body confirmed that it can detect postures deterioration based on the changes in angles formed by the nose, neck, and pelvis. In the future, we would like to further improve the reliability of the posture deterioration detection by combining multiple joint angles for practical use.

References

[1] J. Martinez, R. Hossain, J. Romero, and J. J. Little, "A Simple Yet Effective Baseline for 3d Human Pose Estimation," arXiv preprint arXiv:1705.03098, pp. 1-10, 2017.

[2] W. Yang et al., "3D Human Pose Estimation in the Wild by Adversarial Learning," 2018 IEEE/CVF Conference on Computer Vision and Pattern Recognition, pp. 5255-5264, 2018.

[3] O.D.A. Prima et al., "Evaluation of Joint Range of Motion Measured by Vision Cameras," International Journal on Advances in Life Sciences , 11, 3 & 4, pp. 128-137, 2019.

[4] C. Ionescu, D. Papava, V. Olaru, and C. Sminchisescu, "Human3.6M: Large Scale Datasets and Predictive Methods for 3D Human Sensing in Natural Environments," IEEE Transactions on Pattern Analysis and Machine Intelligence, 36, 7, pp. 1-15, 2014.

[5] K. He, X. Zhang, S. Ren, and J. Sun, "Delving Deep into Rectifiers: Surpassing Human-Level Performance on Imagenet Classification," Proceedings of the IEEE International Conference on Computer Vision, pp. 1026-1034, 2015.

[6] A. Newell, K. Yang, and J. Deng, "Stacked Hourglass Networks for Human Pose Estimation," Lecture Notes in Computer Science, 9912 LNCS, pp. 483–499, 2016.

Gaze Calibration of Eye Trackers for Head-Mounted Displays Using Eye-Frontalization Process

Katsuyoshi Hotta, Oky Dicky Ardiansyah Prima,
Graduate School of Software and Information Science
Iwate Prefectural University
Takizawa, Japan
e-mail: prima@iwate-pu.ac.jp, g236q004@s.iwate-pu.ac.jp

Takashi Imabuchi

Japan Atomic Energy Agency
Fukushima, Japan
e-mail: imabuchi.takashi@jaea.go.jp

Abstract— **In order to simplify the gaze calibration required for gaze measurement in Virtual Reality Head-Mounted Displays (VR-HMDs), we proposed a new gaze calibration method that combines eye-frontalization and single-point calibration. Deep Convolutional Neural Network was used for the frontalization of the eyes. Our method enables to estimate the gaze coordinates on the screen as soon as the user puts on the VR-HMD and to improve the accuracy of gaze measurement due to slight changes in the positional relationship between the VR-HMD and the face by updating the parameters of the gaze calibration. The accuracy of the proposed method was about 5 degrees for both eyes.**

Keywords-VR; EyeTracking; Gaze Calibration; Gaze redirection.

I. INTRODUCTION

Gaze information is essential in research and technology using visual processing. This information provides insight into the characteristics of the object of interest, gazing order, and eye movements. Although we can control the gaze position arbitrarily, eye movements associated with changes in the gaze position are involuntary [1]. Therefore, eye movements are used as response indices to stimuli presented on a screen [2]. In the medical field, these indices are used to determine visual acuity for visual field testing [3] and as input interfaces for Amyotrophic Lateral Sclerosis (ALS) patients [4]. The information of eye movements also has been applied to rendering processing [5] and industrial robot control [6].

Recently, there are Virtual Reality Head-Mounted Displays (VR-HMDs) equipped with eye trackers, such as HTC Vive Pro and Fove, which enable eye tracking in Virtual Reality (VR). These devices have a wider viewing angle of the stimulus and provide more stable eye tracking than non-contact eye trackers. However, if the position of the VR-HMD during the initial gaze calibration and the subsequent position of the VR-HMD relative to the face change, the accuracy of the gaze measurement will be degraded. This problem is caused by its inability to compensate for the gaze point using the corneal reflection, as in non-contact eye trackers. Figure 1 shows the eye tracker mechanism in the VR-HMD and the corneal reflection image indicating that the corneal reflection image cannot be captured correctly when gazing at a wide viewing angle.

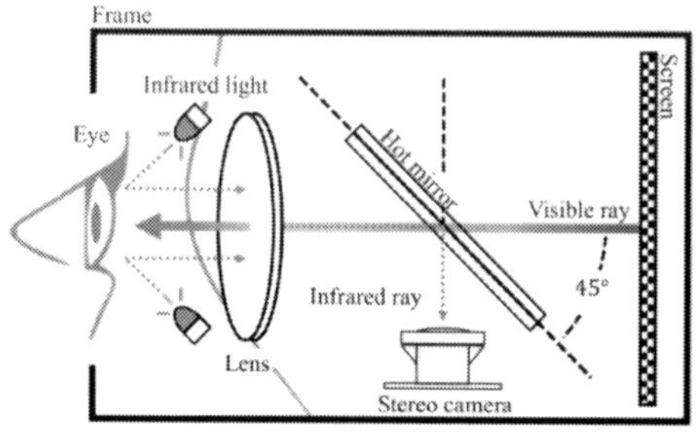

(a) The structure of the VR-HMD eye tracker.

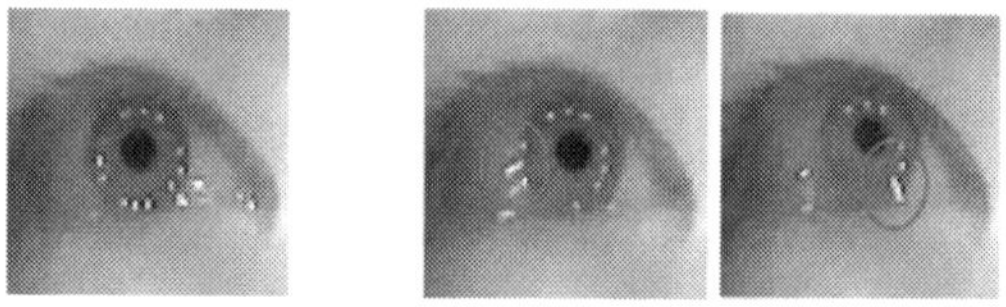

Looking at the front. Looking at an angle.

(b) Corneal reflection images

Figure 1. Eye tracker mounted on the VR-HMD and corneal reflection images captured by the tracker's camera.

In this study, we propose a new gaze calibration method that can compensate the accuracy of gaze measurement for changes in the position of the VR-HMD relative to the face. The proposed method consists of eye-frontalization and single-point calibration. The former involves inferring the image of an eye gazing in the frontal direction from the image of an eye gazing in an arbitrary direction. This method was inspired by the study of gaze redirection [7]. The latter is a higher-order polynomial that connects the gaze direction of the frontal gaze to the direction of the surroundings. This polynomial allows to estimate the gaze point when gazing at the screen (Point of Regard; PoR) without gaze calibration if the image of the eye while looking at the center of the screen can be obtained [8].

Our proposed method will have the following advantages. First, since the single-point calibration can be performed automatically by frontalizing the eyes, the PoR can be obtained instantly after the user wears the VR-HMD. Secondly, the gaze accuracy will not be degraded even if the

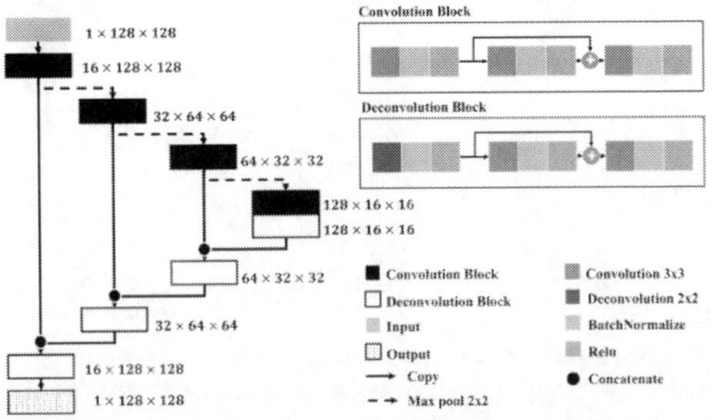

Figure 2. The convolution neural network for the eye-frontalization.

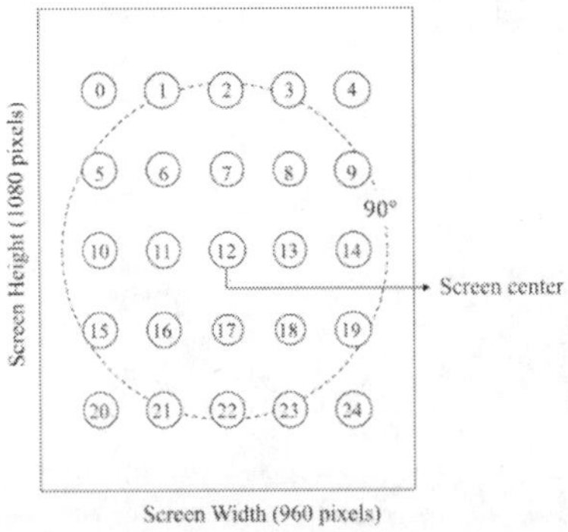

Figure 3. 25 visual targets for data collection in this study.

positional relationship between the VR-HMD and the face changes, thus enabling gaze measurement in long-term.

This paper is organized as follows. In Section II, we describe related work on the frontalization process of eye images. In Section III, we describe the frontalization process and how to obtain the gaze position. In Section IV, we present the validation methods and results on reconstruction error and gaze accuracy. Finally, the results of this research and future perspectives are presented.

II. Related Work

Gaze redirection is a method of modifying the gaze in any direction for a given eye image. This method is an emerging research topic in computer vision and computer graphics, especially in applications such as the generation of eye contact in video conferencing to improve communication.

The initial work on gaze redirection was conducted by Wolf et al. [7]. Their method detects the position of an eye and replaces it with the image of the most similar eye in the dataset, looking straight at the camera. A simple deformable eye model was used to adjust the size and orientation of the eye area. The limitations of this method are that the results depend on the accuracy of the eye region detection and the inability to handle large head movements. Wood et al. [9] proposed a 3D model of a large eye region covering the eyelids and lower eye area to achieve a more natural eye resemblance. The fitting of the 3D model to the eye region was done by minimizing the energy in the similarity of both images and landmarks of the model and the target eye. The disadvantages of this method are the high computational cost that makes it impractical to run in real-time. Instead of using a large 3D model, Isikdogan et al. [10] used the Eye Correction Model Network (ECC-NET) to guide the gaze from the eye region cropped by the facial landmark detector and applied the Generative Adversarial Network (GAN) to the U-Net based network to generate natural-synthetic eye images. Moreover, in order to stabilize the output image, temporal filtering was performed before generating the final output.

TABLE I. Specifications of Experimental Equipment

Device	Element	Specification
VR-HMD Eyetracker	FoV	90°
	Resolution(Mono)	960 × 1080
	Luminance	250cd/m2
	Sampling Rate	240Hz
Desktop PC	CPU	i7-9700K(3.6GHz)
	GPU	RTX2080(8GB)
	Memory	16GB
	OS	Windows 10 64bit(Ver.2004)

III. Proposed Gaze Calibration Method

The proposed gaze calibration method consists of three processes: eye-frontalization, pupil center localization, and PoR calculation.

A. Eye frontalization

We build a convolutional neural network (CNN) that performs eye-frontalization by referring to ECC-Net, which is based on U-Net. This network uses a Pre-Activation Module (PAM) [11], which makes it easier to inherit local features. This not only lowers the computational cost but also allows for a better output of eye images. Figure 2 shows the CNN build for this study. The input eye image is set to 128x128 pixels. L1Loss (absolute error) is used for the loss function of the network, as in previous studies.

B. Getting the pupil center coordinates

Blob searching is commonly used to localize the pupil center coordinates from an eye image. By approximating the shape of the blob with ellipse, the center of the resulting ellipse is treated as the center of the pupil. The brightness of the pupil region, however, is not constant due to eyelashes, eyelid shadows, and Purkinje images. Hotta et al. [8] used Semantic Segmentation (SS) to generate a blob image from an eye image, and then used the Starburst algorithm to find the center of the blob [12]. This method provides a stable

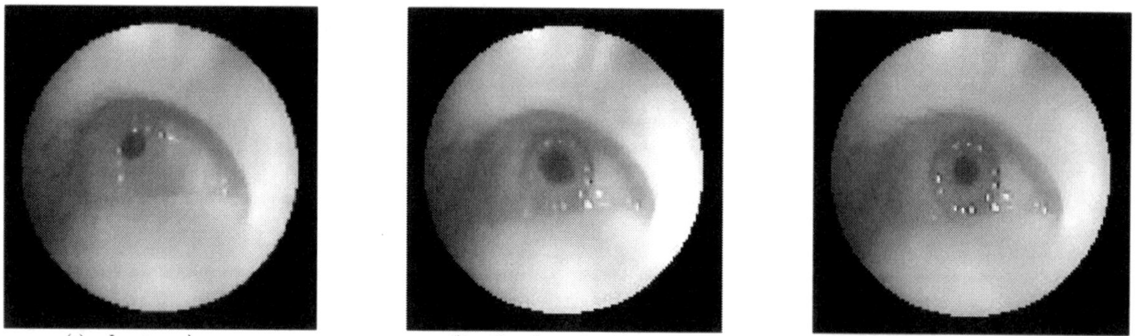

(a) Input eye image (b) Eye-frontalized image (c) The actual eye image gazing the center screen

Figure 4. The input eye image, the resulting frontalized image, and the image of the eye gazing at the center screen.

result of the center of the pupil. In this study, we adopted the method of Hotta et al. [8] to determine the coordinates of the pupil center.

C. PoR calculation

In this study, the PoR is obtained by applying single-point calibration [8] to the coordinates of the pupil center. The relationship between the pupil center coordinates (x, y) after the eye-frontalization and the screen coordinates (u, v) is computed using the following pre-determined n-order polynomial obtained by the single-point calibration experiment.

$$u = \sum_{j=0}^{n} \sum_{k=0}^{j} a_{j,(j-k)} x^j y^{j-k}, \quad v = \sum_{j=0}^{n} \sum_{k=0}^{j} b_{j,(j-k)} x^j y^{j-k} \tag{1}$$

The coefficients $a_{j,(j-k)}$ and $b_{j,(j-k)}$ are obtained by the least-squares method. Finally, the current gazing point (u', v') on the screen (PoR) is calculated by

$$\begin{bmatrix} u' \\ v' \end{bmatrix} = \begin{bmatrix} u' \\ v' \end{bmatrix} + \begin{bmatrix} t_x \\ t_y \end{bmatrix}. \tag{2}$$

Here, (t_x, t_y) represents the horizontal and vertical difference between (u, v) and the screen center coordinate.

IV. EXPERIMENT AND RESULT

36 subjects participated in this experiment. All of the subjects had no visual acuity problems. The experimental setting, data collection, and accuracies of the eye-frontalization and PoR are described as follows.

A. Environmental setting

Table I shows the specifications of VR-HMD and the desktop PC used in the experiment. The software application programs for frontalizing the eye and displaying the stimuli during the experiment were implemented using openFrameworks (C/C++).

B. Data collection and training the CNN

To train the CNN for the frontalization process, we collected images of eyes gazing at the center screen and other locations on the screen. We collected images of all subjects' eyes gazing at 25 visual targets on the screen, as shown in Figure 3. Each subject was requested to fixate on 25 targets three to four times, resulting 4,000 eye images. These images were used for training. Figure 4 shows the input eye image, the resulting frontalized image, and the image of the eye gazing at the center screen.

C. Eye-frontalization accuracy

The accuracy of the frontalized eye image is not determined by the difference in its appearance relative to the image of the eye gazing at the center screen. Instead, both pupil centers are extracted, and the difference in their distances (deviations) is calculated as a measure of eye-frontalization accuracy. Figure 5 shows the distribution of horizontal and vertical deviations for the frontalized left and right eyes. The mean deviation for this experiment was 1.67 ± 0.98 pixels and 1.74 ± 1.04 pixels for the left and right eyes, respectively.

D. Accuracy of PoR

We calculated the PoRs using equation (2) from the images of the eyes when gazing at 24 targets, excluding the center of the screen. The accuracy of PoRs is calculated by

$$Accuracy = \frac{1}{N} \sum_{i=1}^{N} \sqrt{\left(Target_{x_i} - PoR_{x_{i,j}}\right)^2 + \left(Target_{y_i} - PoR_{y_{i,j}}\right)^2} \tag{3}$$

Here, $Target_{x_i}$ and $Target_{y_i}$ are the abscissa and the ordinate of the i-th target, respectively. Likewise, $PoR_{x_{i,j}}$ and $PoR_{y_{i,j}}$ are the abscissa and the ordinate of the i-th PoR, respectively. N is the total number of PoRs.

The accuracy of PoR was 5.07±3.30° and 5.50±3.25° for the left and right eyes, respectively. Figure 6 shows the distribution of PoRs for all subjects. As the accuracy of the typical eye Tracker is approximately 1°, we can say that the accuracy of the proposed method is insufficient. However, it is acceptable for the first attempt of automatic gaze calibration.

V. CONCLUSION AND FUTURE WORK

In this study, we have proposed a new gaze calibration method that combines eye-frontalization and single-point

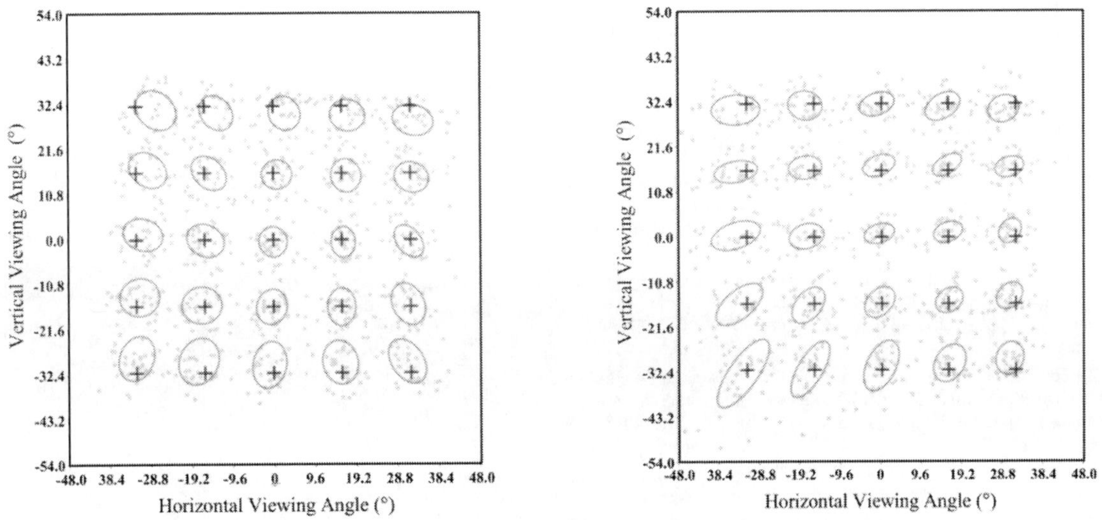

(a) Left eye

(b) Right eye

Figure 6. The distribution of PoRs for all subjects calculated by the proposed method.

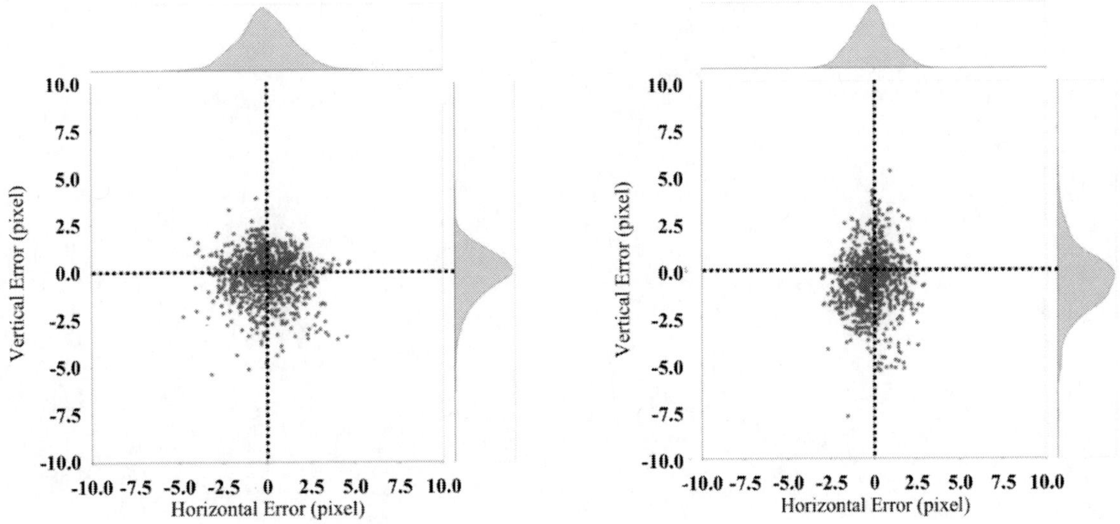

(a) Left eye

(b) Right eye

Figure 5. the distribution of horizontal and vertical deviations for the frontalized left and right eyes.

calibration. Since the accuracy of a typical eye tracker is about 1°, the accuracy of the proposed method can be considered low, but it is within the acceptable range for the first attempt of automatic gaze calibration. The proposed gaze calibration method can be used to compensate for the accuracy of gaze measurement even when the position of the VR-HMD relative to the face changes.

In the future, we will continue to improve the calibration method to make the eye tracking system for head-mounted displays easier to use and more accurate.

REFERENCES

[1] R. Monty and J. W. Senders, Eye Movements and Psychological Processes. Routledge, 2019.

[2] A. Talk, I. Ant-Méndez, and B. Pennefather, "Graded expression of source memory revealed by analysis of gaze direction," PLoS One, vol. 12, pp. e0188727, 2017, doi:10.1371/journal.pone.0188727.

[3] I. Murray et al., "Saccadic Vector Optokinetic Perimetry (SVOP): a novel technique for automated static perimetry in children using eye tracking," Annu Int Conf IEEE Eng Med Biol Soc, pp.3186-3189, 2013, doi:10.1109/EMBC.2013.6610218.

[4] A. Calvo et al., "Eye Tracking Impact on Quality-of-Life of ALS Patients," 11th International Conference on Computers Helping People with Special Needs (ICCHP 2008), Jul. 2008, pp. 70-77.

[5] P. Anjul et al., "Towards foveated rendering for gaze-tracked virtual reality," Association for Computing Machinery, vol. 35, no. 6, pp. 1–12, 2016, doi:10.1145/2980179.2980246.

[6] R. Alessandro, P. Ugo, M. Giorgio, and N. Lorenzo, "A Cartesian 6-DoF Gaze Controller for Humanoid Robots," Proceedings of Robotics: Science and Systems, Jun. 2016, doi: 10.15607/RSS.2016.XII.022.

[7] L. Wolf, Z. Freund, and S. Avidan, "An eye for an eye: A single camera gaze-replacement method," Computer Vision and Pattern Recognition (CVPR 2010), Jun. 2010, pp. 817–824, doi: 10.1109/CVPR.2010.5540133.

[8] K. Hotta, O. D. A. Prima, T. Imabuchi, and M. Kameda, "Development of High-Performance Visual Field Tester Based on The Eye Movement," The Institute of Image Electronics Engineers of Japan, vol. 50, no. 3, 2021.

[9] E. Wood, T. Baltrusaitis, L. P. Morency, P. Robinson, and A. Bulling, "GazeDirector: Fully Articulated Eye Gaze Redirection in Video," Computer Graphics Forum, vol.37, no.2, pp. 217-225, 2018, doi: 10.1111/cgf.13355

[10] I. F. Isikdogan, G. Timo, and M. Gilad, "Eye contact correction using deep neural networks," Winter Conference on Applications of Computer Vision (WACV, 2020), Mar. 2020, doi:10.1109/wacv45572.2020.9093554.

[11] K. He, X. Zhang, S. Ren, and J. Sun ,"Identity Mappings in Deep Residual Networks," European Conference on Computer Vision (ECCV 2016), Mar. 16, vol.9908, ISBN : 978-3-319-46492-3.

[12] D. Li, D. Winfield, and D. J. Parkhurst, "Starburst: A hybrid algorithm for video-based eye tracking combining feature-based and model-based approaches," Computer Society Conference on Computer Vision and Pattern Recognition (CVPR 2005) - Workshops, Sep. 2005, pp. 79–79, doi:10.1109/CVPR.2005.531.

3D Gaze Characteristics in Mixed-Reality Environment

Kenta Kato and Oky Dicky Ardiansyah Prima
Graduate School of Software and Information Science
Iwate Prefectural University
Takizawa, Japan
e-mail: g236s002@s.iwate-pu.ac.jp, prima@iwate-pu.ac.jp

Abstract— **Recently, Head-Mounted Displays (HMDs) have become popular, making it easier to experience Mixed Reality (MR) environments that fuse real and virtual space. With the ability of MR to locate virtual Three-Dimensional (3D) objects in the real space, our 3D perception may also change as this type of 3D experience increases. While there have been studies that measure 3D gaze in either virtual or real space, no studies have discussed how the MR affects 3D gaze. In this study, we developed a See-Through Head-Mounted Display (ST-HMD) to analyze the effect of MR environment on 3D gaze measurement. We conducted experiments in two different physical environments: a room with and without depth cues. Our results showed that there was no significant difference in the measured 3D gaze between rooms with and without depth cues. Experiments of tracking the gaze of a visual target moving from back to front showed that the scanpath of the 3D gaze followed the trajectory of the target's movement.**

Keywords-3D gaze estimation; See-Through Head-Mounted Display; depth perception.

I. INTRODUCTION

The depth cue of binocular disparity was introduced by Wheatstone (1838) for the first time [1]. Since then, the first Three-Dimensional (3D) movie was released in 1922 where anaglyph glasses were used [2]. Later, 3D Televisions were introduced around 2010, but due to low demand, they were discontinued around 2016. 2016 is a symbolic year in which many Virtual Reality Head-Mounted Displays (VR-HMDs) were launched. The HoloLens, a See-Through Head-Mounted Display (ST-HMD), enables users to experience Mixed Reality (MR), new environment and visual representation generated by the fusion of real and virtual world.

Depth perception is classified into binocular cues, which receive three-dimensional sensory information from both eyes, and monocular cues, which are represented in only two dimensions and observed with one eye. Binocular cues include retinal disparity, which exploits parallax and vergence. In contrast, monocular cues include relative size, texture gradient, occlusion, linear perspective, contrast differences, and motion parallax. Displays with head-tracking capability can generate motion parallax to improve the sense of depth.

ST-HMDs are expected to be effectively used in the medical field to allow multiple observers to view 3D medical images in MR [3][4]. These devices are prone to discomfort and fatigue as the viewing time increases [5]. This is due to maintaining the focus of the eye while continuing to gaze at a moving object with vergence eye movements [6]. Analyzing the 3D gaze characteristics of viewers in MR is the key to solve these problems.

3D gaze can be measured using a binocular eye tracker. The representation of 3D gaze can be roughly divided into two categories: the direction of gaze from the eye and the 3D position of the eye-gaze point [7][8]. Recent advances in deep learning technology have made it possible to estimate the gaze direction directly from face images [9], but to estimate the 3D position of the eye-gaze point, a binocular eye tracker with 3D gaze calibration is still needed [10]-[12].

Analyzing 3D gaze in MR environments requires evaluating the relationship between the perceived position of a 3D object and the 3D eye-gaze point of that object. Using the Microsoft HoloLens with a binocular eye tracking, Öney et al. (2020) measured 3D gaze depth during a visual search task of 3D objects in MR environment placed within 1.25m to 5m of the subject. However, their experiment suffered from a significantly large measurement error of more than a meter when the focused object was only 3.5m away from the viewer [13]. To achieve high accuracies in 3D gaze measurements, Kapp et al. (2021) filtered the resulting fixations manually, yielding an approximately 5cm of errors in average within 4.0m measurement distance [14].

In this study, we measure 3D gaze in MR and analyze the influence of the surrounding physical environment on 3D perception cues. Unlike previous studies, we will analyze the characteristics of 3D gaze scan paths of moving targets as well as stationary targets. In addition, we do not perform manual filtering of gaze data to reveal the 3D gaze characteristics in MR environment. The 3D eye tracker with ST-HMD for MR is developed in this study.

The rest of this paper is organized as follows. Section II describes the 3D eye tracker based on ST-HMD developed for this study. Section III explains our experiments in environments with and without depth cues. Section IV describes the results. Finally, Section V presents our conclusion.

II. 3D EYE TRACKER BASED ON ST-HMD

We adopted Moverio (BT-30E, EPSON) as an ST-HMD and a three simultaneous USB camera module (KYT-U030-3NF, KAYETON) to capture images of both eyes and the viewer' scene at the same time. These cameras run at 60Hz. Figure 1 shows our 3D eye tracker. We used Pupil Capture (Pupil Labs), an open-source eye tracking platform, to localize the pupils of both eyes [15]. The software

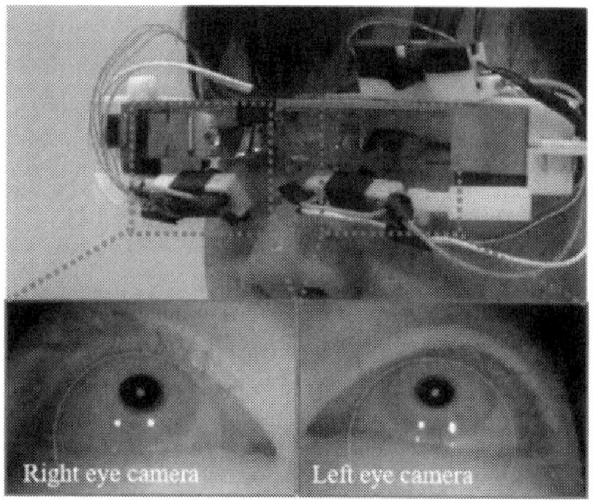

Figure 1. Our 3D eye tracker used in this study.

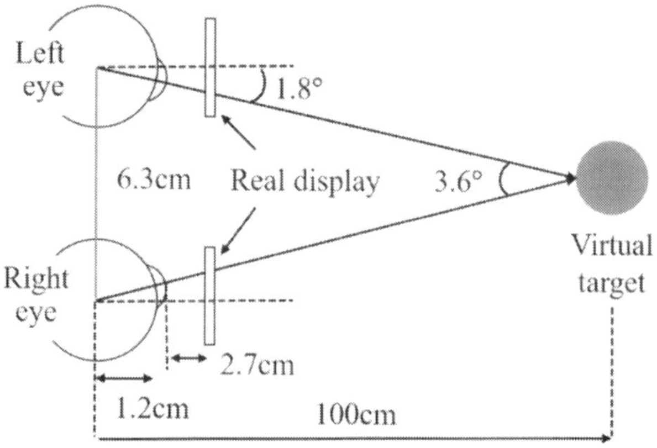

Figure 2. The relationship between both eyes and the virtual

(a) A room with depth cues.

(b) A room without depth cues.

Figure 3. The physical environment for the experiments in this study.

automatically calculates 3D eye vector running from the estimated eyeball center.

A. *Virtual Environments*

The virtual targets used for measurement were created by generating binocular disparity to induce vergence. To determine the disparity, the angle between 3D eye vectors is calculated when gazing at a target placed at a certain distance. The ST-HMD used in this study was designed to allow the user to perceive a virtual screen equivalent to 40-Inch displayed at 250cm. The interpupillary distance was fixed at 6.3cm, which is the average interpupillary distance for Japanese [16]. Figure 2 shows the relationship between both eyes and the virtual target placed 100cm in front of the viewer. We defined the radius of the adult eye to be 1.2cm, as there were no significant differences by gender, race, or age group [17].

B. *3D Gaze Estimation*

To calculate the 3D gaze, a polynomial equation is used to determine the relationship between the gaze direction of both eyes and the 3D position of the gazing target when gazing with vergence eye movement. Let (θ_l, θ_r) represent pitch angles and (φ_l, φ_r) represent yaw angles of eye-sight lines coming from the eye-ball center to the pupil center of the left and right eyes, respectively, the coordinate value of the 3D gaze (G_x, G_y, G_z) is calculated by,

$$G_x = a_1\theta_r^2 + a_2\varphi_r^2 + a_3\theta_l^2 + a_4\varphi_l^2 + a_5\theta_r\varphi_r + a_6\theta_l\varphi_l + a_7\theta_r\theta_l + a_8\theta_r\varphi_l + a_9\theta_l\varphi_r + a_{10}\varphi_r\varphi_l + a_{11}\theta_r + a_{12}\varphi_r + a_{13}\theta_l + a_{14}\varphi_l + a_{15} \quad (1)$$

$$G_y = b_1\theta_r^2 + b_2\varphi_r^2 + b_3\theta_l^2 + b_4\varphi_l^2 + b_5\theta_r\varphi_r + b_6\theta_l\varphi_l + b_7\theta_r\theta_l + b_8\theta_r\varphi_l + b_9\theta_l\varphi_r + b_{10}\varphi_r\varphi_l + b_{11}\theta_r + b_{12}\varphi_r + b_{13}\theta_l + b_{14}\varphi_l + b_{15} \quad (2)$$

$$G_z = c_1\theta_r^2 + c_2\varphi_r^2 + c_3\theta_l^2 + c_4\varphi_l^2 + c_5\theta_r\varphi_r + c_6\theta_l\varphi_l + c_7\theta_r\theta_l + c_8\theta_r\varphi_l + c_9\theta_l\varphi_r + c_{10}\varphi_r\varphi_l + c_{11}\theta_r + c_{12}\varphi_r + c_{13}\theta_l + c_{14}\varphi_l + c_{15}. \quad (3)$$

Coefficients $(a_1 \sim a_{15}, b_1 \sim b_{15}, c_1 \sim c_{15})$ are calculated by the least-squares method based on the correspondence between the gaze direction $(\theta_l, \theta_r, \varphi_l, \varphi_r)$ of each eye and the 3D position of the gazing target obtained by 3D eye calibration.

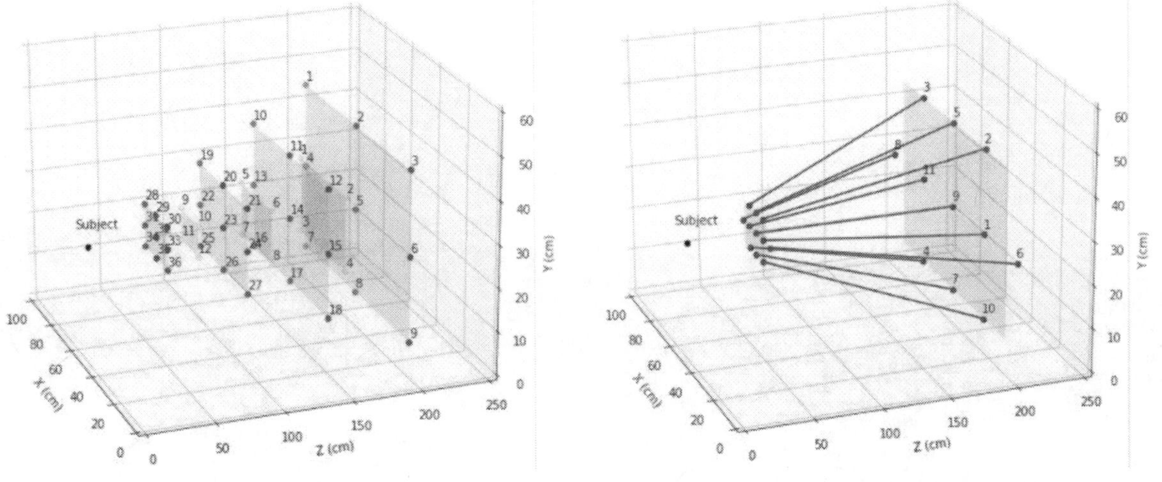

(a) Visual targets for calibrations and validations

(b) The back-to-front moving visual targets

Figure 4. Experimental environment and target placement

III. EXPERIMENT

In this study, we conducted gazing experiments using the following procedure in viewing environments with and without depth cues.

Step 1. The subject is equipped with the ST-HMD and is seated at 250cm from the front wall.

Step 2. 3D eye calibration is performed using 36 visual targets placed in the virtual space. Each target is displayed sequentially in 3s.

Step 3. Another 12 visual targets are displayed to validate the accuracy of the 3D Gaze calibration.

Step 4. The subject is asked to track a visual target that is approaching from 200cm to 50cm in the virtual space. It takes 5 seconds for the target to travel. This tracking is repeated 11 times with targets coming from different directions.

The above procedure is conducted in two different physical environments: a room with and without depth cues. Figure 3 shows the physical environment for the experiments in this study. Here, to place visual targets for 3D gaze calibration, we set up four planes at 50cm intervals in the virtual space ranging from 50 to 200cm from the subject. In each plane, nine targets were placed. For the validation, three surfaces were set up at 50cm intervals at 75 to 175cm from the subject, and four targets were placed on each surface. The eleven back-to-front moving visual targets are used to examine the change in depth of the 3D gaze.

IV. RESULTS

Four subjects (male, mean age 23.3) participated in the experiment. They were tested for visual acuity using a Landolt ring to confirm that their vision achieved 1.0 or better. They were also asked to fill out a questionnaire to confirm that they had no health concerns.

A. Measurement Accuracy

Accuracies for the 2D gaze measurement Acc_{2D} is measured by

$$Acc_{2D} = \sqrt{\frac{1}{n}\sum_{i=1}^{n}\frac{\pi}{180}atan\left(\frac{\sqrt{(T_{xi}-G_{xi})^2+\left(T_{yi}-G_{yi}\right)^2}}{T_{zi}}\right)}, \quad (4)$$

whereas and the 3D gaze measurements Acc_{3D} is measured by

$$Acc_{3D} = \sqrt{\frac{1}{n}\sum_{i=1}^{n}(T_{xi}-G_{xi})^2+\left(T_{yi}-G_{yi}\right)^2+(T_{zi}-G_{zi})^2}. \quad (5)$$

Here, n is the number of targets used for the measurement, T_{xi}, T_{yi}, T_{zi} and G_{xi}, G_{yi}, G_{zi} are the coordinates of the i-th target and the associated eye-gaze points.

Tables I and II show the Acc_{2D} and Acc_{3D} of the resulted calibrations and validations conducted in two physical environments. The accuracy of the 2D gaze calibration is less than 3 degrees, regardless of the experimental environment. This value corresponds to an error of less than 5cm at an area 1m away from the eye. For validations, we observed that the accuracy decreased in the environment with depth cues. On the other hand, there was no significant difference in accuracy for both calibration and validation of the 3D gaze. These accuracies were analyzed with a 2 × 2, depth cues × visual targets, two-way Analysis of Variance (ANOVA). Effects were not found in either the depth cues (F (1, 12) = 0.192, p = .669) or the visual targets (F (1, 12) = 0.192, p = .669; F (1, 12) = 3.497, p = .086). We conducted post-experimental

TABLE I. ACCURACIES FOR THE 2D GAZE MEASUREMENT (°)

Subject	With depth cues		Without depth cues	
	Calibration	*Validation*	*Calibration*	*Validation*
1	2.96	3.06	1.98	2.01
2	3.20	3.34	3.28	2.22
3	2.12	2.10	2.26	3.89
4	3.02	8.75	2.13	1.91
Mean	2.83	4.31	2.41	2.51
Std. Dev.	0.478	3.006	0.589	0.930

TABLE II. ACCURACIES FOR THE 3D GAZE MEASUREMENT (cm)

Subject	With depth cues		Without depth cues	
	Calibration	*Validation*	*Calibration*	*Validation*
1	21.38	24.74	11.48	21.73
2	20.03	23.68	26.65	26.77
3	7.40	11.07	12.85	20.74
4	22.55	34.79	9.78	22.92
Mean	17.84	23.57	15.19	23.04
Std. Dev.	7.037	9.721	7.744	2.643

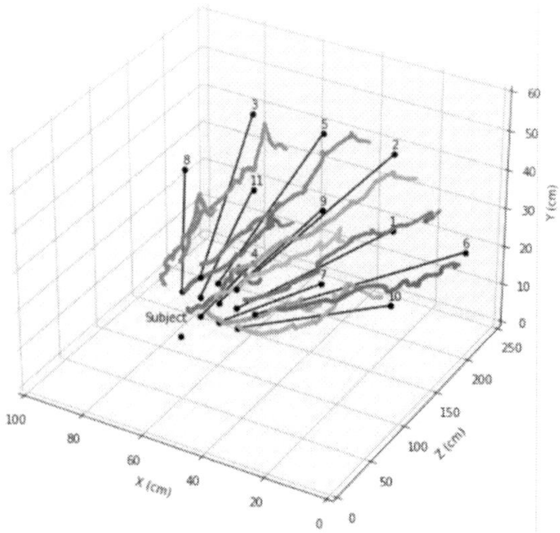

Figure 5. The path along which each target moves and its gazing position.

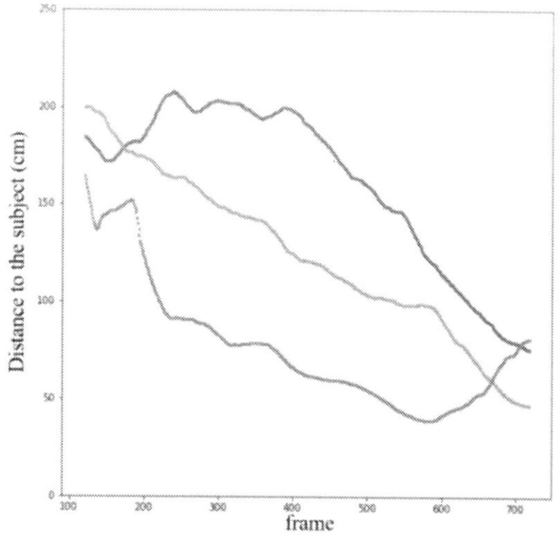

Figure 6. Change in gazing distance with movement of the target.

interviews to find out how the users perceived the 3D visual targets. The post-experimental interview revealed that the fourth subject was unable to converge on some of the validation targets. However, since the accuracy of the calibration was sufficiently high that the data from the subsequent experiments could be used for analysis.

B. Difference between perceived distance and measured distance

After the gaze calibration and validation completed, we performed experiments to track the back-to-front moving visual targets with the gaze. Figure 5 shows a typical of subject's scanpath of 3D gaze while tracking the back-to-front targets. The black line indicates the line connecting the initial position of the target in the front-back direction and the position of the subject's eyes. As can be seen, the scan path of the 3D gaze follows the trajectory of the target motion.

Figure 6 shows the profile of typical 3D scanpaths of #7 back-to-front moving visual target. The orange line shows a downward trend, indicating that the distance measured by the 3D gaze gets shorter as the back-to-front visual target gets closer. This is an ideal result, but not all subjects were able to produce these characteristics of eye movements. Subjects observed that their 3D gaze became unstable when the targets started to move, as shown by the blue and green lines. In the post-experimental interviews, we found that this was due to difficulties in tracking the visual target with proper vergence when it started moving.

V. CONCLUSSION

In this study, we measured and analyzed 3D gaze in MR environments. Our experiments showed that there was no significant difference in the measured 3D gaze between rooms with and without depth cues. This result is consistent with that indicated by Öney et al. (2020), but our setup achieves an average accuracy of less than 25cm, which is about four times better than their measurements [13].

Experiments with tracking the gaze of a visual target moving from back to front showed that the 3D gaze scanpath followed the trajectory of the target's movement. However, in some situations, some users found it difficult to track the visual target with proper vergence when it started moving. In MR environments, it is common for viewers to view 3D contents in motion, thus further analysis of the relationship among viewer movements and 3D gaze characteristics is important.

With the increasing use of VR devices, MR will become more accessible, and there will be more research on the characteristics of vergence eye movements in 3D experiences. In our next experiments, we plan to measure 3D gaze while the subject is moving to investigate the effect of motion parallax.

REFERENCES

[1] K. R. Brooks, "Depth Perception and the History of Three-Dimensional Art: Who Produced the First Stereoscopic Images?" I-Perception, vol. 8, no. 1, pp. 1-22, 2017, doi:10.1177/2041669516680114.

[2] S. Kunić and Z. Šego, "3D Television," Proceedings ELMAR-2011, pp. 127-131, 2011, ISSN: 1334-2630, ISBN: 978-1-61284-949-2.

[3] T. Sielhorst, C. Bichlmeier, S. M. Heining, and N. Navab, "Depth Perception – A Major Issue in Medical AR: Evaluation Study by Twenty Surgeons," Medical image computing and computer-assisted intervention : International Conference on Medical Image Computing and Computer-Assisted Intervention (MICCAI), 9, pp. 364-372, 2006, doi:10.1007/115866565_45.

[4] J. E. Swan, A. Jones, E. Kolstad, M. A. Livingston, and H. S. Smallman, "Egocentric depth judgments in optical, see-through augmented reality," in IEEE Transactions on Visualization and Computer Graphics, vol. 13, no. 3, pp. 429-442, 2007, doi:10.1109/TVCG.2007.1035.

[5] M. Emoto, Y. Nojiri, and F. Okano, "Changes in Fusional Vergence Limit and its Hysteresis after Viewing Stereoscopic TV," Displays, 25, pp. 67-76, doi:10.1016/j.displa.2004.07.001.

[6] T. Shibata, J. Kim, D. M. Hoffman, and M. S. Banks, "Visual discomfort with stereo displays: Effects of viewing distance and direction of vergence-accommodation conflict," Proceedings of SPIE - the International Society for Optical Engineering, vol.7863, pp. 1-9, 2011, doi:10.1117/12.872347.

[7] A. Kacete, R. Séguier, M. Collobert, and J. Royan, "Head Pose Free 3D Gaze Estimation Using RGB-D Camera," ICGIP, SPIE, vol. 10255, 2016, Tokyo, Japan, hal-01393594.

[8] L. Świrski, and N. A. Dodgson, "A fully-automatic, temporal approach to single camera, glint-free 3D eye model fitting," In Proceedings of ECEM 2013.

[9] X. Zhou et al., "Learning a 3D Gaze Estimator With Adaptive Weighted Strategy," in IEEE Access, vol. 8, pp. 82142-82152, 2020, doi: 10.1109/ACCESS.2020.2990685.

[10] S. Weber, R. Schubert, S. Vogt, B. M. Velichkovsky, and S. Pannasch, "Gaze3DFix: Detecting 3D fixations with an ellipsoidal bounding volume," Behavior Research Methods, vol. 50, no. 5, pp. 2004-2015, 2018, doi:10.3758/s13428-017-0969-4.

[11] T. Pfeiffer, M. E. Latoschik, and I. Wachsmuth, "Evaluation of Binocular Eye Trackers and Algorithms for 3D Gaze Interaction in Virtual Reality Environments," JVRB - Journal of Virtual Reality and Broadcasting, vol. 5, no. 16, 2008, doi: 10.20385/1860-2037/5.2008.16.

[12] E. G. Mlot, H. Bahmani, S. Wahl, and E. Kasneci, "3D Gaze Estimation using Eye Vergence," BIOSTEC, pp. 125-131, 2016, doi:10.5220/0005821201250131.

[13] S. Z. Öney et al., "Evaluation of Gaze Depth Estimation from Eye Tracking in Augmented Reality," ACM Symposium on Eye Tracking Research and Applications, pp. 1-5, 2020, doi:10.1145/3379156.3391835.

[14] S. Kapp, M. Barz, S. Mukhametov, D. Sonntag, and J. Kuhn, "ARETT: Augmented Reality Eye Tracking Toolkit for Head Mounted Displays," Sensors, vol. 21, no. 6, 2234, 2021, doi: 10.3390/s21062234

[15] M. Kassner, W. Patera, and A. Bulling, "Pupil: an open source platform for pervasive eye tracking and mobile gaze-based interaction," Proceedings of the 2014 ACM International Joint Conference on Pervasive and Ubiquitous Computing: Adjunct Publication, pp. 1151-1160, 2014, doi:10.1145/2638728.2641695.

[16] M. Kouchi, and M. Mochimaru, "2008: Anthropometric Database of Japanese Head 2001," National Institute of Advanced Industrial Science and Technology, H16PRO-212.

[17] I. Bekerman, P. Gottlieb, and M. Vaiman, "Variations in Eyeball Diameters of the Healthy Adults," Journal of ophthalmology, vol. 2014, 503645, doi:10.1155/2014/503645.

Home Monitoring Service for Critically Ill Children

Edith Maier, Pascale Baer-Baldauf, Claudia Pedron, Ulrich Reimer
Institute for Information and Process Management
Eastern Switzerland University of Applied Sciences
St. Gallen, Switzerland
e-mail: edith.maier@ost.ch, pascale.baer@ost.ch, claudia.pedron@ost.ch, ulrich.reimer@ost.ch

Abstract—**Although the benefits of remote patient monitoring (home monitoring) are widely recognized, there has been no large-scale integration into regular healthcare delivery. The paper describes a home monitoring service aimed at severely ill children. A smart algorithm detects if parameters have exceeded pre-defined thresholds and thus may imply a critical situation. The service has been developed by a team of medical, technological and research partners and is offered by professional nurses employed by an ambulatory care association. The paper describes the technical solution of the monitoring service, its infrastructure, and relevant processes as well as its integration into the healthcare market. The paper finishes with outlining possible future developments that could help promote the uptake of such a service.**

Keywords–healthcare; eHealth; digital health; home care; remote monitoring; support service.

I. INTRODUCTION

Thanks to recent technological advances, such as smart devices, sensors and mobile technologies, the remote monitoring of patients who suffer from chronic disease or who can be taken care of at home instead of being hospitalized has become feasible. Remote patient or home monitoring has come to be an attractive option because healthcare systems are under intense pressure due to an ageing population with chronic diseases and comorbidities, demographic and lifestyle trends, and shortage of healthcare personnel [1][2]. Besides, due to the Corona crisis, there appears to be an increasing willingness to accept and pay for services delivered online, such as teleconsultations, remote monitoring, etc.

Improvements in quality, efficiency, and accessibility of care, especially in rural or remote areas, have been identified as the main benefits of home monitoring. Potential beneficiaries include patients, who can stay at home, but also healthcare personnel, such as physicians or nurses who may monitor the status of their patients continuously. This allows early recognition of health problems or exacerbations (e.g., in the case of obstructive pulmonary disease) as well as monitoring the effect of medication or a therapy plan as well as a patient's adherence to it. Remote monitoring may also encourage patients to take on an active role in the delivery of the service, e.g., by monitoring their vital functions or carrying out necessary interventions on their own.

Despite the known benefits, many remote patient monitoring projects do not develop beyond the initial pilot phase [3][4]. Quite often the initial funding comes from research grants or special digital health initiatives. Not surprisingly, there is an abundance of successful pilot projects, which fail to be introduced into the regular healthcare delivery systems, i.e., the so-called 'first healthcare market', which includes reimbursable products, medicines, and services. Because of the widespread reluctance or inability of patients to pay for telemedicine services, adoption in the *first healthcare market* is a prerequisite for the long-term success and sustainability of a home monitoring service.

To succeed in the first healthcare market, country-specific requirements and characteristics have to be taken into account. For a home monitoring service provider to have its services reimbursed by the health or invalidity insurance, this will usually imply furnishing evidence for the cost-effectiveness, impact, and usefulness of a particular service. According to [5], the *second healthcare market*, i.e., privately financed health services, is primarily helpful as a launching pad for new solution into the primary healthcare market.

In this paper, we present a home monitoring solution that was especially developed for remotely monitoring severely ill children and thus relieving their parents from many nights of interrupted sleep or no sleep at all. Section II provides an overview of related work. Section III describes the methodological approach we have adopted to develop the solution. In Section IV we outline the technical aspects of the monitoring solution and how it has been embedded in an organizational context and a process infrastructure. We especially focus on the challenges that we encountered during the project and how we met them. In Section V we go on to discuss the hurdles, prerequisites, and opportunities of introducing our home monitoring service into the first healthcare market. Finally, Section VI sets out possible future developments to extend the services offered by a home monitoring system. In this section we also present some policy considerations (e.g., reimbursement approaches) that could help promote the uptake of home monitoring services in general.

II. RELATED WORK

Our home monitoring solution has been inspired by a general trend towards digital, and in particular mobile, health systems based on the use of sensors and medical devices [6]. These applications include monitoring systems which collect data either with sensors embedded in smartphones, such as accelerometers, gyroscopes, GPS, microphones or cameras [7] or by means of wearables, such as smartwatches, and are able to measure physiological parameters like heart rate, blood pressure, respiration rate, oxygen saturation, body temperature or electrocardiogram [8].

Reference [6] conducted a comprehensive literature review of remote patient monitoring systems which covered remote monitoring systems based on sensors attached to the body, ambient sensors and systems based on contactless camera-based methods. The authors show that remote monitoring is suitable for a wide range of conditions. Among the most common are those directed at chronic diseases, such as diabetes, the cardiovascular and respiratory systems, fall detection and mobility-related diseases, which are addressed mainly at the elderly, as well as neurological disorders and mental health [9][10].

It appears that only a few monitoring systems address children, in particular, who tend to have special requirements, e.g., with regard to the size of sensors or how these can be attached to the body [11]. In [12] a camera-based monitoring system for hospital environments is discussed, which measures respiration rates and detects apnea using a Kinect camera. Reference [11] proposes a smart monitoring solution based on wearable sensors and a smartphone that continuously monitors a child's activity and vital signs. Whenever the system detects any deviations, it sends an alert to the caregivers (e.g., parents or teachers) as well as the physician in charge. A similar system based on wearable vests is presented by [13].

III. METHODOLOGICAL APPROACH

We relied on a design science approach [14] for the development of the monitoring system. We evaluated the monitoring prototype in various field tests and conducted several iterations, evaluating and refining the prototype and the processes relevant for the remote monitoring solution.

In line with a participatory human-centred design approach, we conducted semi-structured interviews with the nurses who would be in charge of monitoring the sick children in the headquarters of the ambulatory care association for children. The results were analysed and subsequently translated into a set of detailed functional and non-functional requirement specifications for the monitoring application.

We decided early on to use existing devices that had already obtained medical approval. We then examined how these could best be integrated into the application scenario. Apart from medical approval, devices had to meet additional requirements, most importantly providing access to raw, non-aggregated data measured in real-time. Moreover, the option to switch off the video and audio recording of the monitoring devices on the patient side was required so as to ensure people's privacy.

The results from the iterative testing were continuously fed into the further development and adjustments of the various components, be it the adaptation of the thresholds, the alert process, or the graphic user interfaces of the dashboard. The processes relevant to remote monitoring were identified in close collaboration with representatives of the ambulatory care association, validated in the tests and adapted to the findings.

With a view to the later acceptance of our solution in the health care sector it was essential that our monitoring solution should not be inferior to a hospital setting in terms of reliability and accuracy, which is why we conducted an equivalence test in a children's clinic. The test included a comparison of all triggered alerts, a video analysis of both monitors as well as an observation of the test persons by experts.

We furthermore examined different approaches to measure respiration rate, which included impedance pneumography and respiration rate estimation, both from electrocardiogram (ECG) and from audio signals. For the latter we used a microphone attached to the throat, a measure which because of its obtrusiveness, was only applied in the pre-tests. The respiration rate values showed some deviations compared with the manual counting performed on both systems. Besides, we found artefacts in all measurements, which is why we decided to use this parameter just in combination with other parameters to reduce the rate of false alarms.

The hospital test was followed by a test on healthy children in a home setting and further tests in a special home for seriously ill children which is supervised by the ambulatory care association and offers their parents temporary relief from their care responsibilities. Such a scenario comes closest to the home setting, which for both time constraints and ethical considerations could not be implemented during the project period.

IV. DESCRIPTION OF HOME MONITORING SOLUTION

The solution is composed of the following components:

- the medical devices and sensors,
- the data transfer architecture,
- an on-demand audio and video surveillance system.

Since the home monitoring solution must not be inferior to monitoring in a hospital setting, we opted for a multipurpose HEACO monitor which is also used by clinics. It allows to observe physiological parameters including an electrocardiogram, SpO2 (oxygen saturation), PR (pulse), RESP (respiration) and TEMP (temperature).

The need for the audio-visual components only emerged in the course of the project. The nurses involved in the project are employed by an ambulatory care association which initiated the project. Besides, a software development company and a children's hospital participated in the project. The task of the hospital was to make sure that the solution to be developed was on par with the monitoring service in the hospital with regard to reliability, safety and accuracy of measurements. Apart from the technological challenges, the solution had to be embedded in a process landscape and be associated with a detailed organizational plan with regard to authorization, roles and tasks.

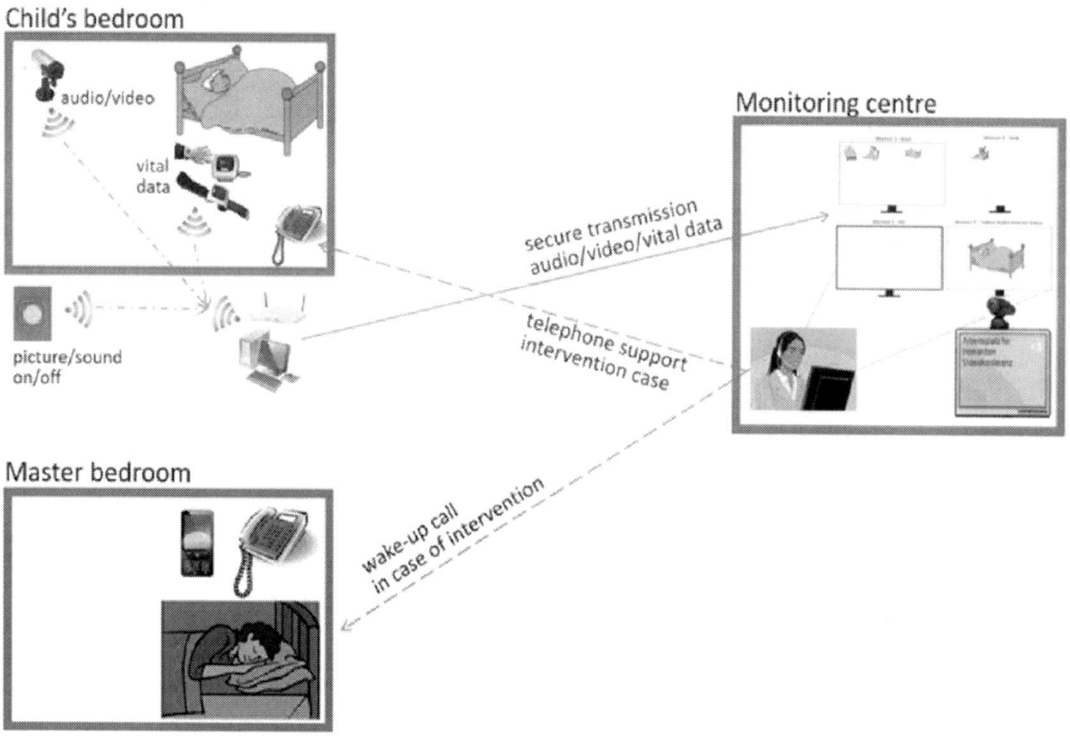

Figure 1. Application scenario for remote monitoring.

For the purpose of monitoring, sensors are attached to the body of the child and the values of the vital parameters are continuously transmitted to a monitoring center based at the headquarters of the ambulatory care association. There, experienced nurses watch the data streams on a dashboard (see Figure 1). Comparable to a traffic lights system, the color red indicates a potential danger or possible emergency, yellow stands for caution and green means 'ok'. If vital signs exceed or fall below the thresholds set by a doctor or the vital parameters are not transmitted, the color turns yellow or red. In the case of a red alarm, the nurse can wake the parents and, if necessary, talk them through the appropriate measures by phone. In severe cases, the professional staff in the monitoring center may contact the ambulance straight away. To prevent false alarms, a video surveillance system enables the nurse to observe the child and judge, for example, if an epileptic seizure is imminent or if the parents should be wakened.

The choice of vital parameters was largely determined by the physician involved in the project and the parameters usually monitored in the hospital, i.e., respiration rate, heart rate and oxygen saturation. These parameters tend to be relevant for the majority of diseases. Besides, they swiftly respond to changes in a patient's condition.

The proposed solution includes a smart algorithm that detects if a certain combination of parameters has exceeded predefined thresholds and thus may imply an emergency.

A. Monitoring infrastructure

The tests carried out with the prototype showed that an infrastructure could be established that is comparable to a stationary monitoring system as used in hospitals. The test in the hospital which involved running two systems in parallel showed that the monitoring system developed for the home setting was not inferior to the hospital setting.

After extensive discussions and testing with the nurses to be in charge, the following quantitative parameters or the monitoring unit were defined: Depending on the case mix and the expected number of alarms, up to 20 patients can be monitored simultaneously. Based on organizational constraints, a child can be placed under surveillance for a maximum of 12 hours. The service is offered from 8pm to 8am, with two active nurses and one nurse on stand-by. The set-up consists of a central dashboard and three individual surveillance stations as depicted in Figure 2.

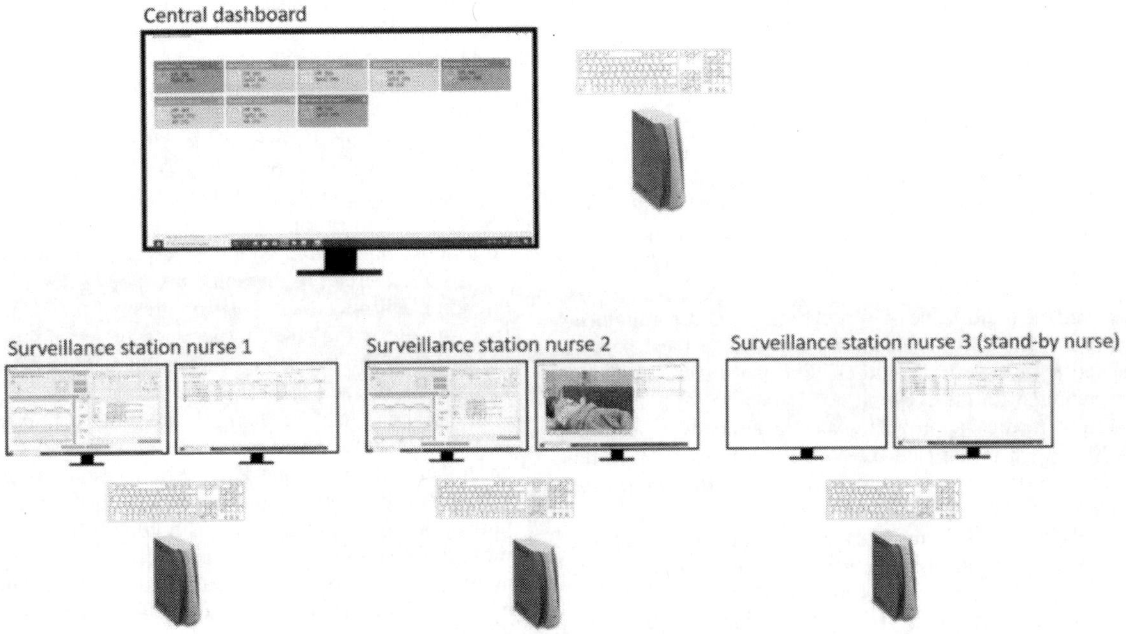

Figure 2. Schematic overview of monitoring unit.

The central dashboard provides an overview of all the patients that are being monitored, giving their names, the most recent values of each vital parameter and the status colors (green, yellow or red). A patient's status color can only be downgraded from red to yellow or to green if the patient has been inspected in one of the surveillance stations. The status of yellow does not trigger an acoustic signal but exhorts the nurse to look at a particular patient more closely. The thresholds for triggering the alarms have to be pre-defined by the physician for each individual patient.

The partner network is especially important in the proposed service: The ambulatory care association for children offers the remote monitoring service. Its specialist nurses operate the monitoring units and monitor patients in real time for up to 12 hours during the night. The same nurses pay regular visits to the families for treatment and basic care services. They have established profound relationships with the families. Another important partner is the paediatrician who defines the thresholds for the vital parameters that are measured and transmitted to the monitoring unit.

Finally, an infrastructure partner provides the monitoring infrastructure and platform, supports the set-up of operation at the patient's home and assures the operability of the monitoring units.

B. Relevant processes

A major task consisted in defining the processes to accompany the implementation of a monitoring system. The processes relevant were identified in close collaboration with representatives of the ambulatory care association, validated in the tests and adapted to the findings. These included authorization and login, the actual monitoring process, informing/alerting parents, alerting emergency services, and contingency plans. It was above all defining the interaction processes between nurses, caregivers at home and emergency staff, e.g., when to issue an alert and whom to alert whenever there was an emergency, that caused lengthy discussions.

The implementation of the planned monitoring solution therefore required developing a detailed *process reference model* that described the relevant processes associated with remote monitoring. A central issue is related to the question what to do if a threshold is exceeded. For example, should the nurse alert the parent immediately or first make sure that it really is an emergency?

The concept of emergency itself proved to be far from unambiguous and triggered extensive discussions in the project team. It is closely related to the definition of thresholds for the various parameters, which tend to differ considerably between individuals. It also involved defining the number of children that a professional nurse might monitor simultaneously.

The main challenge consisted in reducing the high number of alarms per child per night. In the beginning, we registered up to 250 alarms per child and night. We were able to halve that number by adjusting the algorithm. By fine-tuning the thresholds, we further managed to bring the number down to ten to 35 alarms per child and night. In addition to that we decided to introduce audio-visual inspection to be carried out by the professional nurses. The inspection serves as a preliminary check so as not to obliviate the whole purpose of the project, which is facilitating a good night's sleep for the parents. Finding the right balance between safety or caution and running the risk of missing an emergency proved to be a major challenge.

V. Introduction of Home Monitoring into Regular Healthcare Delivery

The overall purpose of the service offered is to support the family caregivers (parents) during the night hours as the responsibility of caring for a severely ill child often comes at the cost of lack of sleep as well as physical and psychological strain, which may lead to chronic stress and eventually physical and mental break-down [15]. This has been confirmed by the findings of a research report commissioned by the Swiss Ministry of Social Insurance, which shows that one in five caregivers never or only rarely enjoys a good night's sleep [16].

Given the importance of introducing our home monitoring service into the first healthcare market, the business partners attached a great deal of importance to the organizational aspects required for the eventual reimbursement of the service. Therefore, medical approval by the relevant body is necessary (the Swissmed in the case of Switzerland). To achieve this, evidence must be furnished to prove the usefulness, cost-effectiveness and efficiency of the solution.

Usefulness: Every year, the ambulatory care association of Eastern Switzerland is responsible for around one hundred mentally and physically handicapped or chronically ill infants, children and young adults. Its clients also include prematurely born children as well as sick, injured, convalescent children and those recovering after surgery. The purpose of the home monitoring service is both preventive and therapeutic: it should detect exacerbations of the child's state of health on the one hand and enable early interventions on the other. Providing more restful nights and guidance by a nurse in an emergency helps to increase parents' endurance as a caregiver. The remote monitoring service also has a therapeutic effect since care at home tends to contribute to the recovery and enhance the quality of life of children with severe illnesses or disabilities [17][18][19].

Cost-effectiveness: Home monitoring is a service offered by the ambulatory care association in addition to its on-site services and replaces night watches by nurses as well as the children's parents. It can therefore be considered an extension to its current portfolio. Due to their 1:1 setting, night watches are very expensive because of the high cost of labour. Night watches are recognized by the Swiss Ministry of Social Insurance and thus a tariff item number exists. This tariff item number is a prerequisite for invoicing the service. However, a remote monitoring service currently has no tariff item number and therefore cannot be invoiced.

This is why it is crucial that the Ministry of Social Insurance as well as the health insurance introduce, such a tariff item number. Otherwise, the business model for the home monitoring service will not be sustainable in the long run. As a result of the Corona crisis, chances are good that this might happen in the near future.

Efficiency: In the preliminary business plan, the service provider calculated the savings due to the productivity increases as a result of remote monitoring. The new home monitoring service allows the simultaneous monitoring of up to 20 patients by three nurses (two active nurses and one nurse on stand-by) in the central monitoring unit, whilst for traditional night watches, only one patient can be monitored at a time. As a result, the fully inclusive fee for the remote monitoring service turns out to be about five times cheaper than an on-site night watch performed by a nurse.

Given the fact that severely ill children usually need to receive intensive care to monitor their health state, the costs tend to be even higher. The experience of the ambulatory care association shows that without their services, children in need of care have to be transferred back to hospital twice as often. In short, as the home monitoring service would substantially support the parents, readmissions to hospital could be prevented.

A further effect of the home monitoring service is the reduction of the secondary costs of stress of the caring parents. In a survey of the stress experience of parents of severely handicapped children, 86.6% of parents stated that taking care of their child costs them a lot of strength and energy [20]. Avoiding or at least reducing these health care costs, which may also include loss of income from paid work, is a further argument in favour of introducing home monitoring.

There is further potential in the more efficient use of scarce nursing staff. At present, nurses take over night watches to provide relief and a good night of sleep to the parents. This activity is very monotonous and according to the ambulatory care association for children, quite a few nurses would therefore be very interested in taking over the remote monitoring of several children instead. Thus, a home monitoring service would make more efficient use of the short supply of nursing staff [21].

It is above all the technical partner who may be interested in exploring new business opportunities, e.g., by extending its services to other partners, user groups or healthcare providers.

VI. Conclusion and Future Work

From a medical point of view, continuous remote monitoring has proven to be feasible and, if implemented properly, is equivalent or not inferior to monitoring in a hospital environment. Audio-visual inspection in the case of possible emergencies crystallized as a central element in remote monitoring, which had not been considered in the original study design.

Furthermore, it has been shown that during the first nights close supervision by the attending physician is desirable so as to be able to adjust individual threshold values.

In the future, it is conceivable that the home monitoring service might evolve in the direction of self-learning systems for decision support as well as automated generation of recommendations for interventions. However, to be accepted by consumers, the algorithms and thus the decision-making process behind the recommendations would have to be made transparent.

Besides, the analysis of the captured vital data may lead to the development of a decision-making aid for adjusting treatment or medication. The storage of vital data is a central component of the prototype and the basis for analysing the test data. However, any such tool will have to conform with data protection legislation, which implies anonymizing the data and/or obtaining the consent of both parents and the child to use their data.

As far as the long-term sustainability of the service is concerned, home care agencies appear to be a promising avenue to pursue as also pointed out by [22]. The business model associated with home care agencies is largely dependent on fixed case-rate reimbursement. This could also be applied to our solution provided that home monitoring is officially recognised. This might involve maintaining patient contact, assessing needs, counselling caregivers while reducing staff travel time.

A major challenge faced by home care agencies is finding a way to share in the much larger savings they generate, e.g., by taking care of chronic care patients or sick children at home instead of hospitalizing them. However, the recent introduction of Diagnosis-Related Group (DRG) system in Switzerland might help persuade hospitals as well as insurance companies to enter into partnerships. Because of the DRG system, hospitals are interested in making sure that care at home is of such a quality that the likelihood for a patient to be readmitted within 30 days is minimized. Any expenses incurred by a 'premature' readmission will have to be covered by the same flat amount, i.e., the hospital is not entitled to any additional payments.

ACKNOWLEDGMENT

The work described in this paper was funded by the Swiss Commission for Technology and Innovation (CTI).

REFERENCES

[1] M. Mars, "Telemedicine and advances in urban and rural healthcare delivery in Africa," *Progress in cardiovascular diseases*, vol. 56, no. 3, pp. 326–335, 2013, doi: 10.1016/j.pcad.2013.10.006.

[2] V. Saliba et al., "Telemedicine across borders: a systematic review of factors that hinder or support implementation," *International Journal of Medical Informatics*, vol. 81, no. 12, pp. 793–809, 2012, doi: 10.1016/j.ijmedinf.2012.08.003.

[3] M. Gupta, L. Chotard, Ó. Ingþórsson, J. Basto, and I. Borges, "Health@home – an e-service model for disease prevention and healthcare in the home," in *Electronic Healthcare: eHealth 2008*, D. Weerasinghe, Ed., 2009, pp. 17–24, doi: 10.1007/978-3-642-00413-1_3.

[4] A. A. M. Spil and B. Kijl, "E-health Business Models: from pilot project to successful deployment," *IBIMA business review*, vol. 1, pp. 55–66, 2009.

[5] F. Rappert, "Business Model Design im zweiten Gesundheitsmarkt/Business model design in the second healthcare market," master thesis, Freie Universität Berlin, Berlin, 2018.

[6] L. P. Malasinghe, N. Ramzan, and K. Dahal, "Remote patient monitoring: a comprehensive study," *Journal of Ambient Intelligence and Humanized Computing*, vol. 10, no. 1, pp. 57–76, 2019, doi: 10.1007/s12652-017-0598-x.

[7] N. D. Lane et al., "A survey of mobile phone sensing," *IEEE Communications Magazine*, vol. 48, no. 9, pp. 140–150, 2010, doi: 10.1109/MCOM.2010.5560598.

[8] A. Pantelopoulos and N. G. Bourbakis, "A survey on wearable sensor-based systems for health monitoring and prognosis," *IEEE Transactions on Systems, Man, and Cybernetics, Part C (Applications and Reviews)*, vol. 40, no. 1, pp. 1–12, 2010, doi: 10.1109/TSMCC.2009.2032660.

[9] Y. Guo and G. Bai, "A general architecture for developing a sustainable elderly care e-health system," *International Journal of Information Technology and Business Management (JITBM)*, vol. 27, no. 1, pp. 95–101, 2014, doi: 10.1109/ICSSSM.2011.5959443.

[10] M. P. Rajasekaran, S. Radhakrishnan, and P. Subbaraj, "Elderly patient monitoring system using a wireless sensor network," *Telemedicine and e-Health*, vol. 15, no. 1, pp. 73–79, 2009, doi: 10.1089/tmj.2008.0056.

[11] S. Sendra, L. Parra, J. Lloret, and J. Tomás, "Smart system for children's chronic illness monitoring," *Information Fusion*, vol. 40, pp. 76–86, 2018, doi: 10.1016/j.inffus.2017.06.002.

[12] A. Al-Naji, K. Gibson, S.-H. Lee, and J. Chahl, "Real time apnoea monitoring of children using the Microsoft Kinect sensor: a pilot study," *Sensors*, vol. 17, no. 2, p. 286, 2017, doi: 10.3390/s17020286.

[13] R. Jansi, R. Amutha, and S. Radha, "Remote monitoring of children with chronic illness using wearable vest," in *Telemedicine Technologies*, H. D. Jude and V. E. Balas, Eds., 2019, 121–137.

[14] A. R. Hevner, S. T. March, J. Park, and S. Ram, "Design science in information systems research," *MIS quarterly*, pp. 75–105, 2004, doi: 10.2307/25148625.

[15] R. Schulz and P. R. Sherwood, "Physical and mental health effects of familiy caregivers," *Journal of Social Work Education*, vol. 44, sup3, pp. 105–113, 2008, doi: 10.5175/JSWE.2008.773247702.

[16] M. Gehrig, J. Guggisberg, and I. Graf, "Wohn- und Betreuungssituation von Personen mit Hilflosenentschädigung der IV: Eine Bestandsaufnahme im Kontext der Massnahmen der 4. IVG-Revision/Housing and care situation of persons with disability helplessness allowance: a stocktaking in the context of the measures of the 4th IVG revision," Bundesamt für Sozialversicherungen, Beiträge zur sozialen Sicherheit 2/2013, 2013. [Online]. Available: https://www.bsv.admin.ch/bsv/de/home/publikationen-und-service/forschung/forschungspublikationen.exturl.html?lang=de&lnr=02/13 [retrieved June, 2021]

[17] A. Bagust, A. Haycox, S. A. Sartain, M. J. Maxwell, and P. Todd, "Economic evaluation of an acute paediatric hospital at home clinical trial," *Archives of disease in childhood*, vol. 87, no. 6, pp. 489–492, 2002, doi: 10.1136/adc.87.6.489.

[18] P. Revill, P. Ryan, A. McNamara, and C. Normand, "A cost and outcomes analysis of alternative models of care for young children with severe disabilities in Ireland," *Alter*, vol. 7, no. 4, pp. 260–274, 2013, doi: 10.1016/j.alter.2013.04.002.

[19] P. Sloper, "Models of service support for parents of disabled children. What do we know? What do we need to know?," *Child: care, health and development*, vol. 25, no. 2, pp. 85–99, 1999.

[20] F. Peterander and O. Speck, "Subjektive Belastungen. Mütter schwerbehinderter Kinder in der Frühförderung/Subjective stress. Mothers of severely disabled children in early intervention," *Geistige Behinderung*, vol. 2, pp. 95–107, 1995.

[21] A. A. Abera, G. H. Mengesha, and P. F. Musa, "Assessment of Ethiopian health facilities readiness for implementation of telemedicine," *Communications of the Association for Information Systems*, vol. 34, no. 1, p. 67, 2014, doi: 10.17705/1CAIS.03465.

[22] M. J. Coye, A. Haselkorn, and S. DeMello, "Remote patient management: technology-enabled innovation and evolving business models for chronic disease care," *Health Affairs*, vol. 28, no. 1, pp. 126–135, 2009, doi: 10.1377/hlthaff.28.1.126.

Use of Medical Teleconsultations During the COVID-19 Pandemic in Poland - Preliminary Results

Urszula Grata-Borkowska
Department of Family Medicine
Wroclaw Medical University
Wroclaw, Poland
e-mail: urszula.grata-borkowska@umed.wroc.pl

Mateusz Sobieski
Department of Family Medicine
Wroclaw Medical University
Wroclaw, Poland
e-mail: mateusz.sobieski@student.umed.wroc.pl

JarosławDrobnik
Department of Public Health
Wroclaw Medical University
Wroclaw, Poland
e-mail: jaroslaw.drobnik@umed.wroc.pl

Ewa Fabich
University Teaching Hospital
Wroclaw Medical University
Wroclaw, Poland
e-mail: efabich@usk.wroc.pl

Maria Magdalena Bujnowska-Fedak
Department of Family Medicine
Wroclaw Medical University
Wroclaw, Poland
e-mail: maria.bujnowska-fedak@umed.wroc.pl

Abstract — **The frequency of teleconsultation by Polish family doctors increased significantly during the COVID-19 pandemic. This study aims to assess the usefulness of teleconsultation, its advantages and disadvantages, as well as its efficiency and credibility. Preliminary results indicate a fairly high efficiency and credibility of teleconsultation.**

Keywords-televisits; primary care; teleconsultations; COVID-19 pandemic; e-health; advantages and disadvantages.

I. INTRODUCTION

E-health services in the form of medical teleconsultations make contact with healthcare professionals easier and more convenient than ever before. Tele-visits, also known as teleconsultations, allow patients to see a doctor from anywhere: on a smartphone, tablet or computer. Virtual physician appointments can save doctors and patients time and avoid crowded waiting rooms, which significantly reduces the risk of infection while waiting for an appointment. This is especially important during COVID-19 pandemic, since when digital access to medical databases and various forms of e-health services has increased significantly [1]. Doctors can conduct a medical interview during tele-visits. In addition, they can refer the patient to laboratory tests and other additional tests; they can make e-prescriptions, e-referrals to specialist doctors or to a hospital ward and issue a required sick-leave. In the period of the coronavirus pandemic, teleconsultation has quickly become one of the basic work tools for primary care physicians (PCPs) in Poland.

Preliminary research conducted among Polish patients (N=1021) regarding their opinions on teleconsultation give ambiguous results - 49% (N=500) of Poles admitted that during the pandemic had a telephone contact with a doctor, and 5% (N=51) - online contact. As many as 58% (N=592) of respondents who used a telephone contact with a doctor were satisfied with such a form of consultation.

Poles are rather not positively focused to telemedicine. Only 25% (N=255) of the respondents prefer contact telephone or online, and three-quarters of the respondents are of the opinion that teleconsultation does not replace direct contact with the doctor. 63% (N=643) of Poles claim that during pandemic patients with other ailments than COVID-19 were treated inadequately. Regardless of the age of respondents, almost half acknowledges that they do not believe in telemedicine, and consider online or telephone visits as completely ineffective [2].

In March 2021, by the ordinance of the Polish government, the rules of the widespread use of teleconsultations have been narrowed due to reported cases of misuse in this area. Nowadays teleconsultations cannot be conducted when it concerns the child under 6 years of age, in case of suspicion of neoplastic disease, or significant exacerbation of a chronic disease.

The aim of the study was to assess the usefulness and suitability of medical teleconsultation as a valuable working tool for primary care physicians in the COVID-19 pandemic. It was important to assess benefits and risks of this form of family doctor's work.

The following research hypotheses were made:

1. For most primary care physicians, teleconsultation is the main working tool.

2. Most family doctors will use teleconsultation in their work after the pandemic COVID-19.

3. According to family doctors, teleconsultations are an effective and reliable tool in everyday work.

II. MATERIALS AND METHODS

The research was based on the specially designed e-questionnaire which was distributed in the online form among all primary care doctors – members of Polish Society of Family Medicine. In this preliminary study, a questionnaire was sent to the 828 family doctors from Lower Silesia District in Poland. 219 (N=219) of them completed the questionnaire, which allowed to achieve a 26% response rate.

The anonymous survey used in the study consisted of a metric including age, gender and any additional specializations performed by the respondent, and 10 questions regarding the respondent's attitude towards teleconsultations. Among 219 respondents, 24% accounted for men (N = 52) and 76 % for women (N = 167); the respondents were in the age between 25-60 years (60%; N=131 in age range 25-49, 40%; N=88 in 41-60). All respondents were doctors; 96% (N=210) of them were family physicians.

III. RESULTS

Over 70% of respondents (N=153) prefer teleconsultation as their primary tool for working in the coronavirus pandemic. 46% (N=101) of the respondents used teleconsultation as a work tool before the COVID-19 pandemic. Almost 90% (N=197) of respondents indicated that the main advantage of teleconsultation is limiting the possibility of the spread of the COVID-19 pandemic. Only 4% (N=8) indicated the possibility of giving advice to more people at the same time.

The most frequently reported disadvantage of teleconsultation is the inability to examine the patient personally and reliably verify his condition - as many as 80% (N=175) of respondents encounter this problem. 10% (N=22) of respondents complain about unreliable information provided by patients themselves. Only 1% (N=2) believe that teleconsultation has no disadvantages.

Most Polish family doctors intend to use teleconsultation after the pandemic is over - as many as 97% (N=212) of doctors expressed a positive attitude towards continuing work using this method, of which 54% (N=118) of respondents intend to use teleconsultation frequently, and 43% (N=94) - occasionally. One percent (N = 2) do not intend to use teleconsultation at all, and 2% (N=4) have no opinion on it.

A frequently reported problem by doctors is the inability to confirm the identity of the person being teleconsulted. Despite this, the vast majority (65%) (N=142) of physicians consider it sufficient to ask for their name and surname. On the other hand, over 20% (N=44) of respondents use a request for a PESEL number as a form of verification, which is the national identification number used in Poland that identifies just one person and cannot be changed to another one.

Significant differences emerged in terms of time spent teleconsultation - 38% (N=83) report that this form of contact they take less time than a traditional visit, 44% (N=96) - as much as 18% (N=39) - more than a personal visit to the patient.

The doctors' assessments of the effectiveness and credibility of teleconsultation are summarized in Figure 1. Teleconsultation was assessed on a scale from 1 to 10, where 1 - the lowest and 10 - the highest efficiency/ credibility. On average, teleconsultation efficiency was rated as 7.22, credibility - as 6.74, which means that doctors believe that teleconsultations have good properties as a work tool. A lower value of credibility than efficiency may result from the problems reported by them in the survey - the inability to identify the patient, the inability to physically examine the patient and simulating symptoms by patients in order to receive sick leave or other treatment, adequate according to patients.

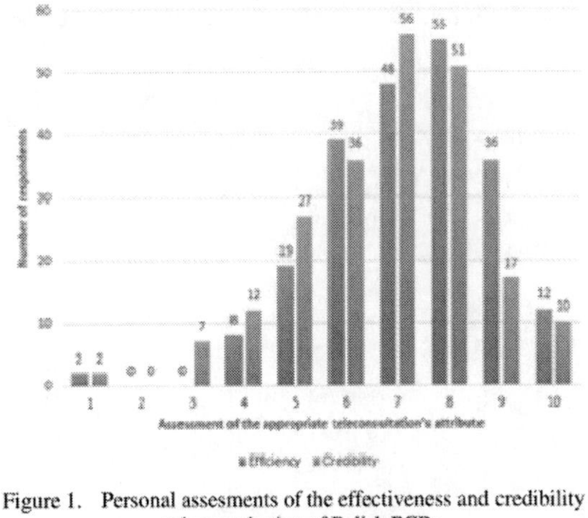

Figure 1. Personal assesments of the effectiveness and credibility of teleconsultation of Polish PCPs.

IV. Discussion

Along with the development of technology, the universal affordability and availability of basic telemedicine tools are growing. The COVID-19 outbreak dramatically changed the situation of the patients and the family physicians. Despite the fact that televisits are a relatively new concept to most people, they are important working tool for epidemic reason. Telemedicine has advantages and disadvantages, however, the most important advantage during COVID-19 pandemic that it is a safe way to receive medical advice [3] [4].

Other Polish surveys also indicate benefits of televisits during the pandemic. Over 90% of the surveyed patients had got a good medical advice during teleconsultations according to a report of a patient satisfaction conducted by the Ministry of Health in cooperation with the National Health Fund and national consultant with the field of Family Medicine [5]. Telemedicine has made it possible to maintain contact between patients and the health care workers in the context of maximum complexity both in Poland and other countries [5][6].

V. Conclusions

Despite the fact that a large percentage of family doctors had the opportunity to use teleconsultation before the COVID-19 pandemic, the inability to conduct traditional visits forced their use by most doctors. According to the respondents, limiting the possibility of spreading COVID-19 infection is the main advantage of teleconsultation. Despite this, most Polish family doctors intend to continue teleconsulting even after the pandemic ends. On the other hand, the main risk is the inability to examine the patient and reliably verify his health. However in their opinion, teleconsultation is characterized by good effectiveness and credibility. The main problem may be the discrepancy between the needs of patients and doctors - the first to use this form of consultation are often satisfied; however, most Poles still prefer stationary visits. Changing this preference will require physicians to adapt to patients' needs more closely and to try to circumvent the limitations of telemedicine.

References

[1] J. Vidal-Alaball et al, "Telemedicine in the face of the COVID-19 pandemic," Atencion Primaria, vol. 52(6), pp. 418-422, Jun-Jul 2020, doi:10.1016/j.aprim.2020.04.003

[2] ARC Rynek i Opinia. Poles reluctant to telemedicine [*Polacy niechętni telemedycynie*]. [Online, in Polish]. Available from: http://arc.com.pl/polacy-niechetni-telemedycynie/ 2021.04.27

[3] U. Pinar et al, Preliminary assessment of patient and physician satisfaction with the use of teleconsultation in urology during the COVID-19 pandemic. World J Urol. 2020 Sep 9:1–6. doi: 10.1007/s00345-020-03432-4. Epub ahead of print. PMID: 32909174; PMCID: PMC7480659.

[4] N.M. Hjelm. Benefits and drawbacks of telemedicine. J Telemed Telecare. 2005;11(2):60-70. doi: 10.1258/1357633053499886. PMID: 15829049.

[5] Report on the satisfaction survey of patients using teleconsultation with a primary health care doctor in the period of COVID-19 pandemic. [Raport z badania satysfakcji pacjentów korzystających z teleporad u lekarza podstawowej opieki zdrowotnej w okresie epidemii COVID-19]. [Online, in Polish]. Available from:

https://www.nfz.gov.pl/download/gfx/nfz/pl/defaultaktualnosci/370/7788/1/raport_-_teleporady_u_lekarza_poz.pdf 2021.06.14.

[6] O.Solans,"Characteristics of Citizens and Their Use of Teleconsultations in Primary Care in the Catalan Public Health System Before and During the COVID-19 Pandemic: Retrospective Descriptive Cross-sectional Study" J Med Internet Res 2021;23(5):e28629

Blind People's Navigation Improvements Using Crowdsourcing

Darius Plikynas
Department of Business Technologies and Entrepreneurship
Vilnius Gediminas Technical University
Vilnius, Lithuania
email: darius.plikynas@vgtu.lt

Audrius Indriulionis
Department of Business Technologies and Entrepreneurship
Vilnius Gediminas Technical University
Vilnius, Lithuania
email: audrius.indriulionis@vgtu.lt

***Abstract*-The literature review and survey of the Blind and Severely Visually Impaired (BSVI) people showed that BSVI are using the same general-purpose or specialized social networking means for communication, learning, remote working, leisure, navigation as other people. BSVI oriented text (and image) to voice, tactile feedback, and other specialized mobile apps or software and hardware solutions help in this matter. This paper studies how crowdsourcing (participatory social networking) can improve navigation and orientation capabilities outdoors and indoors, using the computer vision-based Electronic Traveling Aid (ETA) approach. This study gives insights into the high potential of crowdsourcing usage to improve BSVI people's ETA performance. In this regard, this paper delivers a short overview, BSVI survey results, and a description of the prototype, which we are developing to meet BSVI expectations. Provided insights can help researchers and developers to exploit social Web and crowdsourcing opportunities for BSVI computer vision-based ETA navigation improvements.**

Keywords-social networking; electronic travelling aids; computer vision; blind and severely visually impaired; navigation indoors.

I. INTRODUCTION

Admittedly, a wide range of general-purpose social networks, web 2.0 media apps, and other smart ICT (information and communication technology) tools are developed to improve people's daily tasks, including navigation and orientation. Although they are not destined to meet specialized requirements of BSVI people, but some adds make them useful. For instance, text (and image) to voice, tactile feedback, and other additional enabling software and hardware solutions are helpful for this matter. However, complexity and abundance of features pose a significant challenge for BSVI persons. According to Raufi et al. [1], the volumes of information together with data from social networks confuse BSVI users. In this way, web 2.0 social networks do not guarantee specialized digital content accessibility for BSVI users [2]. Some more focused approaches are in demand.

In general, BSVI users are actively involved in social networks [3]-[5]. More than 90 % of BSVI persons actively use one or more general-purpose social networking sites, such as Facebook, Twitter, LinkedIn, Instagram, and Snapchat [3][4], and [6]-[8]. However, only a few social networking platforms are specifically oriented for BSVI users. For instance, BSVI surveys reveal that social networking apps are among five most popular mobile apps [11]. The majority of BSVI people - who use social media - choose Facebook social networks [4][9], and [10]. The usage of Twitter was also unusually high, assuming that its simple, text-based interface is more accessible to the screen readers [4].

Next to the general-purpose social networks, BSVI people frequently use apps specifically designed for them to accomplish daily activities. However, N. Griffin-Shirley et al. emphasizes that persons with visual impairments would like to see both improvements in existing apps and new apps [11]. Below, we give an example of some of the most popular navigation apps used for path planning, navigation, and obstacle avoidance [12][13], and [3].

For instance, Walky Talkie helps blind people in navigation, providing real-time haptic feedback [14]. However, the accuracy based on the in-built GPS is low. The vOICe for Android application maps live camera views to soundscapes, providing the visually impaired with augmented reality-based navigation support (see The vOICe for Android - Apps on Google Play). Ariadne GPS works where Google Maps are available (see www.ariadnegps.eu/en). In addition to navigation functions, it also enables users to navigate in large buildings by pre-programming locations. BlindSquare provides information to visually impaired users about their surroundings. From a social networking perspective, BlindSquare is closely linked to FourSquare as it collects information about the user's environment from FourSquare [14].

The above mentioned cases and some other navigation apps are mostly based on the pre-developed navigational information, but do not provide a real-life support, user experience-centric approaches, and participatory Web 2.0 social networking. On the contrary, there are other real-life social apps such as Be my eyes, which enable access to a network of sighted volunteers and company representatives who are ready to provide real-time visual assistance for the orientation, navigation and other tasks at hand [15].

In Section 2, we briefly provide a glimpse of BSVI people's survey results concerning their navigation and social networking needs and expectations. In Section 3, we share some insights concerning navigation and orientation for ETA enhancements using participatory Web 2.0 advantages. Conclusions are provided in Section 4.

II. BSVI People's Survey: Social Networking and Navigation Needs

To define more precisely BSVI persons' social networking needs and expectations concerning navigation help, we conducted a survey and semi-structured interview of blind people of various ages. In total, 78 EU located BSVI persons' responses were analyzed, of which 25 were identified as blind experts (10+ years of experience or active interest in using ETAs for the blind). In the survey, some questions (out of 40 questions in total) concerned ETA navigation functionalities, and others dealt with social networking approaches.

For instance, the question "Are you using the assistance of volunteers over electronic means?" (see Figure 1) revealed that only 12 % of all 78 respondents use such assistance. It indicates that the electronic assistance level is currently deficient, bearing in mind the high potential of social networks, web 2.0 media apps, smartphones, and other ICT (information and communication technology) tools. In other words, it points to the lack of enabling real-time, user-friendly, experience–centric, and participatory Web 2.0 technologies.

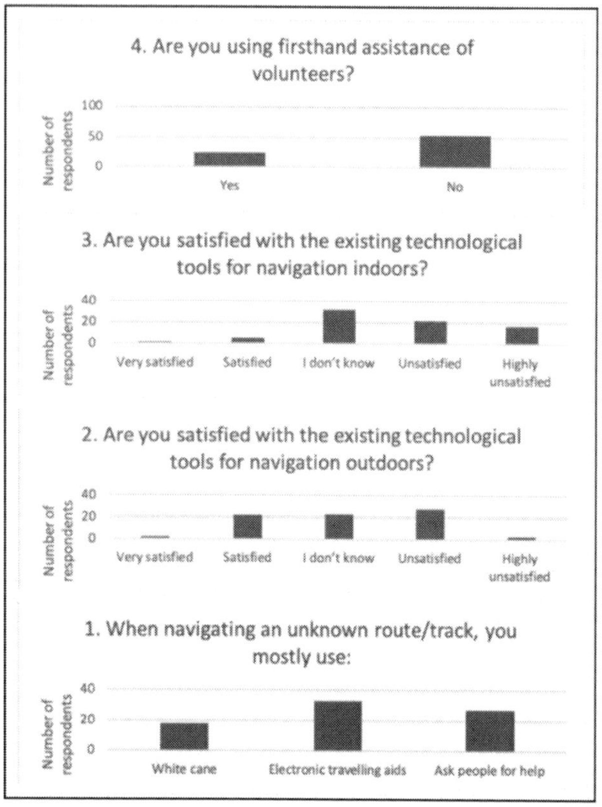

Figure 1. Survey responses from 78 BSVI persons. The order of the questions is piled up from the bottom to the top.

BSVI experts also provided their answers, see Figure 2. They were interviewed concerning the usage of smartphone apps and Web portals. For instance, the questions, such as "What smartphone apps, web portals, and social networks do you know, and which of them do you use to communicate with sighted people?" or "What smartphone apps, web portals and social networks, specifically designed for the blind, do you know?" revealed that most popular apps and social networks used by BSVI people are Facebook, Twitter, LinkedIn, and Snapchat. Besides, BSVI people use Telegram, Youtube, Facetime, Google hangouts, Whatsapp, Skype, Viber, Messenger, Zello, MySpace, Tinder, TeamTalk, Eskimi apps. Only 5 out of 25 BSVI experts mentioned apps or websites specifically designed for the blind: Bee my eyes, Telelight – an accessible telegram client, Voreil, Talking Communities, FourSquare/BlindSquare, Playroom, Applevis.com, Elvis, blindhelp.net, Blindbargens, ACB network, RNIB. It indicates wide variety of smartphone apps and Web portals often used for everyday tasks.

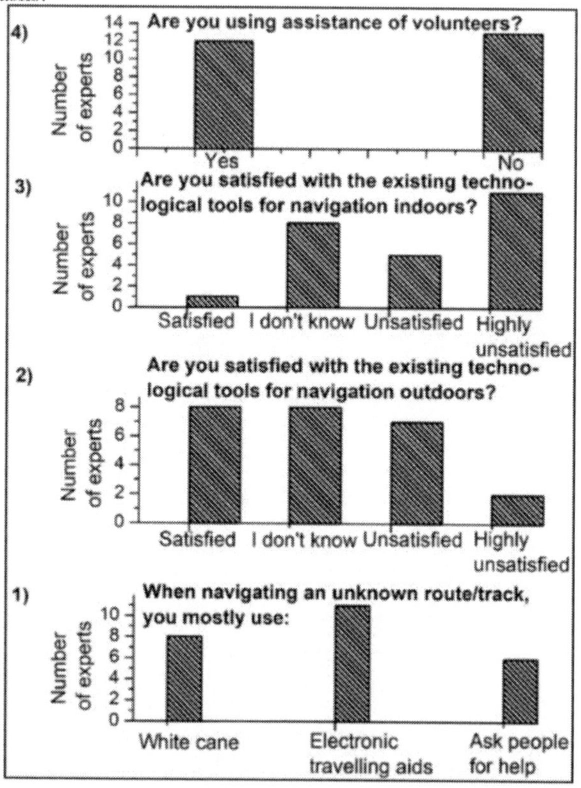

Figure 2. Survey responses from 25 BSVI experts.

Next, we turned our attention to the more specific questions regarding social networking tools used for navigation. For instance, "Are you familiar with social networking tools that support sharing of navigation information (directions) between the blind and/or sighted volunteers?". Surprisingly, most of the BSVI respondents do not know such tools and only a few mentioned What's app, Be my eyes, and Google Groups "Eyes-free group". Thus, social networking tools used for navigation are not very popular. During additional interviews, we figured out a few more details. For instance, social networking tools are (i) in

English mostly; do not operate in other national languages, (ii) casual voluntaries are not accustomed to deal well with the specific BSVI problems, and (iii) applied technology is not specialized enough to tailor real-time and high-quality help.

In another question, "Would you be willing to pay for the functionality of an electronic travel aid listed below? How much?" we identified value-added and monetary estimation of each ETA functionality (however, we present here not the price itself but relative estimates). About 80 % of the first ranked ETA needs included navigation and orientation functionalities, such as recognition of stairs, elevators, doors, navigation directions, assistance to return to a specific location, and so on. Out of the whole price for all twenty chosen ETA functionalities, BSVI were willing to pay 18.3% for outdoor navigation; 12.9% for indoor navigation; 12.5% for recognition of textual and numerical information' 8.4% for recognition of stairs, lifts/elevators, doors, passages and pavements/sidewalks; 6% for information about products with BAR and QR codes; 5.8% for assistance of remote volunteers to interpret sophisticated surroundings in the mother tongue; 3.1% for ability (through social networking) to record, store and reuse outdoor navigation information; 2.7% for ability (through social networking) to record, store and reuse indoor navigation information; 1.9% for ability to share and exchange outdoor navigation directions through a specially designed social network; 1.8% for ability to share and exchange indoor navigation directions through a specially designed social network, and so on. In sum, around 32% out of the total price, BSVI were willing to pay for ETA functionalities, which can be substantially enhanced using participatory Web 2.0 social networking.

These were just a few exemplary questions. However, based on the entire survey analysis, we find out some perspective niche of research and development in the field of navigational ETA solutions. Based on these insights, we made some inferences regarding a combination of modern enabling technologies, which can be successfully employed. In the next section, we give an exemplary case.

III. Machine Vision-Based Navigation Enhancements Using Participatory Social Networking

In this section, we will share a few participatory Web 2.0 social networking ideas, which could enhance BSVI navigation capabilities for traveling, shopping, and other everyday mobility tasks. For instance, in the case of real-time assistance and guidance' for traveling routes, the primary objective of the specialized mobile app's with wearable services is to assist a visually impaired or blind user in navigating from the chosen point A to point B using reliable directions given from an online community. In this case, a phone with wearable service would be able to (i) stream live video to a crowd server (Social Navigation Networking Services) of sighted users through internet/WiFi, (ii) receive real-time feedback through assistance and guidance instructions. For instance, SoNavNet is designed for connected users of the social network to share navigation information with the intent of providing more personalized navigation methods, routes based on member experience rather than the shortest distance [16]. SoNavNet is based on the experience-based approach - through communication (using online social media) and collaboration (sharing and exchanging experiences), BSVIs can find suitable routes both n outdoors and indoors that can meet their specific needs and preferences. SoNavNet, as an online social navigation network system, facilitate sharing and exchanging experiences on Points Of Interest (POIs), Routes Of Interest (ROIs), and Areas Of Interest (AOIs) [16].

The authors in [17] designed a Tales4Us platform to promote creativity, collaboration, and learning process, for a BVSI and other communities to share their shopping stories through a specialized social network. The application has such major functionalities: (i) the user can play other users' shopping stories, (ii) users can record new stories and share it with the community.

In the case of "Seeing-eye person" proposed in [18], a crowdsourcing approach enables multimedia data sharing and services for the BSVI navigation. The goal of this work is to provide user-accessible crowd services (uniquely tailored for visually impaired), flexible (with friendly HCI and APIs for the ease of plugging in new apps to motivate online volunteers for their services), and efficient (near real-time response, and a balanced workload between mobile phone, the back end system, and the different types of users).

The authors in [19] dedicate their general-purpose social navigation approach for any users, including BSVI people with impaired mobility. The system allows sharing knowledge between them, reviewing an existent place freely, or uploading new ones to the global database, improving the application content. The ParticipAct infrastructure, implementing calls to different external API services, as geocoding, localization, routing calculation, and POI entities download, enabling a new set of functionalities. However, the system does not include data quality support in the sense of automatic filtering-out of erroneous inputs, as (possibly) fake entities.

In our ETA research, we also use the onboard sensors of a smartphone (iPhone or Android Phone), such as a camera, compass, GPS, and accelerometer, to assist the navigation of a blind user, see Figure 3. The primary function of mobile computing is to stream the video and other sensory information to the crowd server so that volunteers can use the information to provide service. We plan further to tailor these techniques in order to address the unique challenges in crowd-assisted navigation, such as the smoothness and reliability of visual labeling of routes recorded by volunteers, and the contextual information of video frames.

We infer that some of the navigational instructions might also be adapted from the machine vision algorithms that provide direction information [20]. It lays the potential to aggregate all the available instructions into a single one that will be returned to the blind user. We need to consider the expertise, reliability, and reputation (low reputation might indicate noisily, or even malicious volunteer, which we want to filter) in the data aggregation process. Besides, we also need to consider the synchronicity of data coming from different volunteers and the frequency of instructional updates.

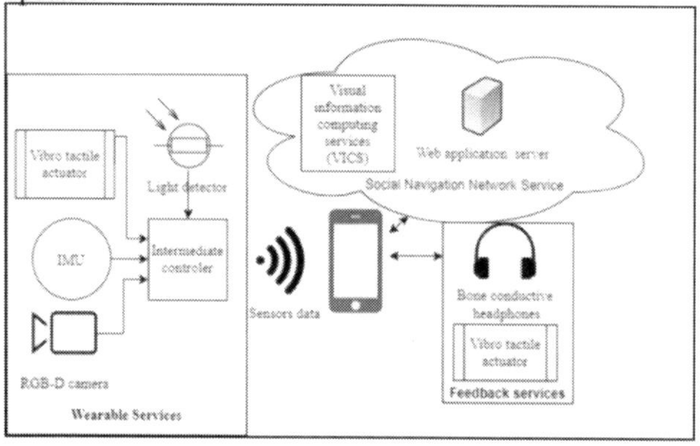

Figure 3. Social navigation network services enabled by mobile device for BSVI

The seamless integration of AI-based vision algorithms in the crowdsourced social networking solutions can provide additional feedback. Vision-based algorithms can be tested for accuracy against live information from human volunteers. Along with the on-line process and data aggregation, an offline analysis will in turn help better tailor context-aware human-computer interfaces and further improve the online analysis tasks.

Based on the survey results and above mentioned considerations, we are reporting an interim result - a prototype of a wearable system configured to help as an offline and online web-crowd assisted decision support system for BSVI people when orientating and navigating in indoor environments (for instance, public institutions, schools, hospitals, airports, stores, and other buildings). Admittedly, there is a lack of feasible indoor navigational solutions that would work well without GPS signal and prearranged infrastructural indoor installations (such as WI-FI routers, beamers, RFID tags). Our survey of blind experts has shown that after outdoor navigation, the second most demanded and not satisfied need concerns ETA solutions for indoor navigation and orientation [12] and [13]. We figured out that BSVI persons need ETA for orientation and navigation in unfamiliar indoor environments to detect and recognize desired indoor destinations such as rooms, WC, staircases, elevators, avoiding obstacles on their way. In Figure 4, we depicted key guidelines for the ETA system's interface with the BSVI person indoors.

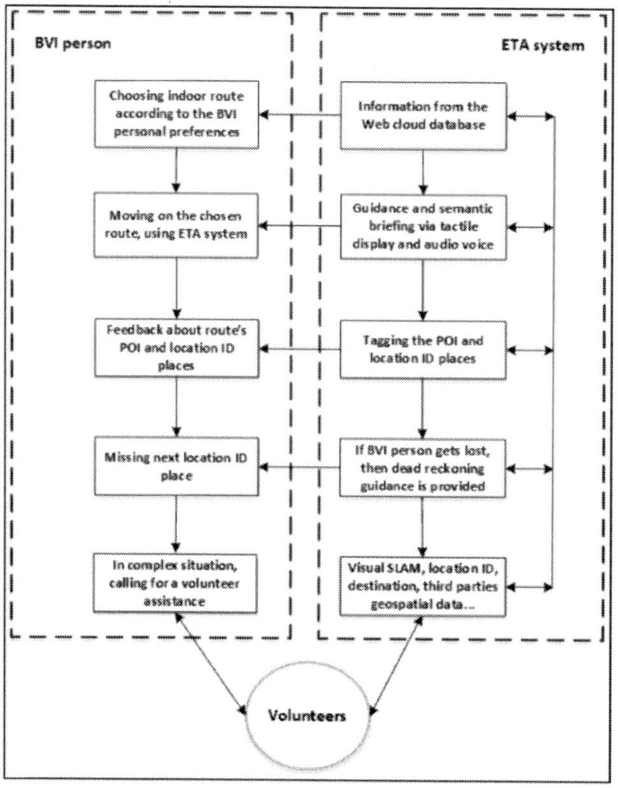

Figure 4. Web crowd (volunteers) assisted method for indoor routing enhancement and optimization, using ETA system functionality

The presented system is a compound technology of innovatively adapted hardware devices like the 3D ToF IR camera, RGB camera, specially designed tactile display with EMG sensors, bone-conducting earphones, controller, and

IMU, GPS, light detector, compass sensors. GSM communication can be implemented as a stand-alone device or smartphone that can work as an intermediate processing device. Passive sensors passively collect environmental data, whereas, active sensor like 3D ToF IR camera emits IR light to estimate distances to the objects. Multi-sensory data is used to (i) find needed objects, (ii) locate obstacles, and (iii) infer users' location in an indoor environment in order to help navigate. The devices and sensors observe the environment in real-time and send data via the controller to the machine learning processing, where features' extraction, object recognition, and data storage occur in the web cloud database server, see Figure 4. The prototype integrates devices and interfaces using modern technologies and methods from machine learning and computational vision domain.

From the point of view of the end-user, this prototype distinguishes among other related wearable indoor navigational ETA novelties in the sense of a) intelligent user interface integrity based on unique tactile display and audio instructions, b) hands-free intuitive control interface using EMG (electromyography), c) comfortable user-orientated headband design, d) machine learning-based real-time guidance, e) web-crowd assistance while mapping indoor navigational routes and solving problematic situations on the way.

For efficient indoor navigational performance, the presented ETA system is used in three consequently interconnected modalities: (i) Web crowd assistance when volunteers go through buildings and gather step-by-step indoor routes' information that is processed in the web cloud server and stored in the online DB; (ii) BSVI usage of web cloud DB indoor routes when they need guided navigational assistance; (iii) in complex indoor situations (such as getting lost, encountering unexpected obstacles and situations), the BSVI ETA system's multisensory data stream can be used in real-time to get voice-guided help from volunteers familiar with the particular route or building.

IV. Conclusions

This paper studied how crowdsourcing (participatory social networking) can improve navigation and orientation capabilities outdoors and indoors, using the computer vision-based ETA (electronic traveling aid) approach. This study gave insights into the high potential of crowdsourcing usage to improve BSVI people's ETA performance. In this regard, this paper delivered a short overview, BSVI survey results, and a description of the prototype, which we are developing to meet BSVI expectations. Provided insights can help researchers and developers to exploit social Web and crowdsourcing opportunities for BSVI computer vision-based ETA navigation improvements. More specifically, semi-structured survey revealed a clear lack of participatory Web 2.0 social networking usage for the navigation and orientation outdoors and indoors. From one side, it is related with the lack of BSVI people's trust and confidence in the corresponding ETA technological solutions. From another side, it is related with the lack of enabling real-time, user-friendly, user experience–centric, and participatory Web 2.0 social networking technologies.

We found that current research in the area of online social navigation network systems, facilitate sharing and exchanging experiences on POIs, ROIs, and AOIs. We inferred that some of the navigational instructions might also be adapted from the machine vision algorithms that provide direction information [20]. The seamless integration of AI-based vision algorithms in the crowdsourced social networking solutions can provide additional feedback. Vision-based algorithms can be tested for accuracy against live information from human volunteers.

In summary, participatory Web 2.0 social networking systems can enable the integration of smart algorithms together with BSVI and sighted people's best experiences while traveling, navigating, and orientating in outdoor and indoor environments. It helps to build and continuously update real-time metrics of reachable and unreachable POIs to the effect that routes could be averaged, erroneous routes eliminated, and user experience-based optimal solutions found using various optimization approaches (like min-max entropy calculation) [21]-[23]. In this way, crowdsourced navigation social platforms with real-time video streaming and analysis get advantages that none stand-alone navigation systems could ever achieve [9]. It is appealing; even there are still several unresolved tasks like data reliability, integrity, synchronicity, cross-platform compatibility [10] and [24].

The presented ETA system uses crowdsourcing when volunteers go through buildings and gather step-by-step indoor routes' visual and other sensory information that is processed, using machine learning algorithms, in the web cloud server and stored for the BSVI usage in the web cloud DB.

In the presented paper, we provided some framework and concepts of ETA enhancement for indoor guided navigation, using outsourcing of routes mapping. Next, we are step by step developing a wearable ETA device supplemented with the social networking interface, wherein sighted users take an active part in mapping indoor routes and helping BSVI persons in complex situations.

In time, a participatory web 2.0 social networking platform – something like a worldwide "Visiopedia - with big, labeled, crowdsourced, almost real-time updated, and publicly available outdoor and indoor navigational database for BSVI could emerge. It would enable much more efficient and reliable use of AI-based learning algorithms [12] and [13].

Acknowledgment

This project has received funding from European Regional Development Fund (project No 01.2.2-LMT-K-718-01-0060) under grant agreement with the Research Council of Lithuania (LMTLT).

REFERENCES

[1] B. Raufi, M. Ferati, X. Zenuni, J. Ajdari, I. F. Ismaili, "Methods and techniques of adaptive web accessibility for the blind and visually impaired," Procedia - Social and Behavioral Sciences, vol. 195(2015), pp. 1999 – 2007, 2015.

[2] M. Ferati, B. Raufi, A. Kurti, B. Vogel, "Accessibility requirements for blind and visually impaired in a regional context: An exploratory study. 2nd International Workshop on Usability and Accessibility Focused," Requirements Engineering, UsARE 2014, pp. 13-16, 2014.

[3] M. Elbes, A. Al-Fuqaha, "Design of a social collaboration and precise localization services for the blind and visually impaired," Procedia Computer Science, vol. 21(2013), pp. 282 – 291, 2013.

[4] E.Brady, Y. Zhong, M. R. Morris, J. P. Bigham, "Investigating the appropriateness of social network question asking as a resource for blind users," Proceedings of the 2013 conference on Computer supported cooperative work, San Antonio, Texas, USA, pp. 1225-1236, February 23–27, 2013.

[5] J. Brinkley, N. Tabrizi, "A Desktop Usability Evaluation of the Facebook Mobile Interface using the JAWS Screen Reader with Blind Users," Proceedings of the Human Factors and Ergonomics Society 2017 Annual Meeting, pp. 828 – 832, 2017.

[6] Sh. Wu, L. Adamic, "Visually impaired users on an online social network," Proceedings of the SIGCHI Conference on Human Factors in Computing Systems, pp. 3133-3142, 2014.

[7] V. Voykinska, Sh. Azenkot, Sh. Wu, G. Leshed, "How blind people interact with visual content on social networking services," Proceedings of the 2016 conference on Computer supported cooperative work, San Francisco, CA, USA, pp. 1584–1595, February 27-March 02, 2016.

[8] S. Qiu, J. Hu, G. W. M. Rauterberg, "Mobile social media for the blind: Preliminary observations," Proceedings of the International Conference on Enabling Access for Persons with Visual Impairment (ICEAPVI), Athens, Greece, pp. 152-156, 12-14 February 2015.

[9] B. Della Líbera, C. Jurberg, "Teenagers with visual impairment and new media: A world without barriers," British Journal of Visual Impairment, vol. 35(3), pp. 247–256, 2017.

[10] I. Tjostheim, I. Solheim, K. S. Fuglerud, "The importance of peers for visually impaired users of social media," Proceedings of the 6th IASTED International Conference on Human-Computer Interaction, pp. 23–30, 2011.

[11] N. Griffin-Shirley, D. R. Banda, P. M. Ajuwon, J. Cheon, J. Lee, H. R. Park, S. N. Lyngdoh, "A survey on the use of mobile applications for people who are visually impaired," Journal of Visual Impairment & Blindness, vol. 111(4), pp. 307-323, 2017.

[12] D. Plikynas, A. Žvironas, A. Budrionis, A., M. Gudauskis, "Indoor Navigation Systems for Visually Impaired Persons: Mapping the Features of Existing Technologies to User Needs". Sensors, 20(3), 636, 2020.

[13] D. Plikynas, A. Žvironas, M. Gudauskis, A. Budrionis, "Research Advances of Indoor Navigation for the Blind People: A Brief Review of Technological Instrumentation," IEEE Instrumentation & Measurement Magazine, vol. 23(4), pp. 22-32, 2020.

[14] Á. Csapó, G. Wersényi, H. Nagy, T. Stockman, "A survey of assistive technologies and applications for blind users on mobile platforms: A review and foundation for research," Journal on Multimodal User Interfaces, vol. 9 (4), pp. 275-286, 2015.

[15] A. Mauro, K. Wolf, A. Brock, and N. Henze. "Remote assistance for blind users in daily life: A survey about be my eyes." In Proceedings of the 9th ACM International Conference on PErvasive Technologies Related to Assistive Environments, pp. 1-2. 2016.

[16] E. Sheepy, S. Salenikovich. "Technological Support for Mobility and Orientation Training: Development of a smartphone Navigation Aid." Proceedings of World Conference on E-Learning in Corporate, Government, Healthcare, and Higher Education 2013 (pp. 975-980). T. Bastiaens & G. Marks (Eds.), 2013.

[17] N. Karlsson, E. Di Bernardo, J. Ostrowski, L. Goncalves, P. Pirjanian, M.E. Munich, "The vSLAM Algorithm for Robust Localization and Mapping," Proceedings of the 2005 IEEE International Conference on Robotics and Automation (ICRA 2005), 2005. pp. 24-29. doi: 10.1109/ROBOT.2005.1570091

[18] B. Zimmerman, A. Ozcelik, D. Roongpiboonsopit, "SoNavNet: A Framework for Social Navigation Networks". In International Workshop on Location Based Social Networks (LBSN'09). Seattle, WA, November 3-6, 2009.

[19] S. Vitek, M. Klima, L. Husnik, D. Spirk, "New possibilities for blind people navigation". In proceedings of 2011 International Conference on Applied Electronics (AE): 1-4, 2011.

[20] G. Olmschenk, C. Yang, Z. Zhu, H. Tong and W. H. Seiple, "Mobile Crowd Assisted Navigation for the Visually Impaired," 2015 IEEE 12th Intl Conf on Ubiquitous Intelligence and Computing (UIC-ATC-ScalCom), Beijing, pp. 324-327, 2015.

[21] B. K. Kanan, N. Kothari, C. Gnegy, H. Gedaway, M. F. Dias, M. B. Dias, "Localization, Route Planning, and Smartphone Interface for Indoor Navigation". In Cooperative Robots and Sensor Networks, pp. 39-59. Springer Berlin Heidelberg, 2014.

[22] H. K. Gedawy, "Designing an Interface and Path Translator for a Smart Phone-Based Indoor Navigation System for Visually Impaired Users". Masters Thesis, Carnegie Mellon University, 2011.

[23] K. Nisarg, B. Kannan, E. D. Glasgwow, M. B. Dias, "Robust indoor localization on a commercial smart phone". Procedia Computer Science 10: 1114-1120, 2012.

[24] E. Pissaloux, R. Velazquez (Eds.), "Mobility of Visually Impaired People: Fundamentals and ICT Assistive Technologies". Springer, 2018.

Speech and Language Relearning for Stroke Patients- Understanding User Needs for Technology Enhancement

Awais Ahmad
Computer and Systems Science,
Mid Sweden University
Östersund, Sweden
awais.ahmad@miun.se

Peter Mozelius
Computer and Systems Science
Mid Sweden University
Östersund, Sweden
Peter.mozelius@miun.se

Karin Ahlin
Computer and Systems Science
Mid Sweden University
Östersund, Sweden
Karin.ahlin@miun.se

Ali Hassan Sodhro
Computer and Systems Science
Mid Sweden University
Östersund, Sweden
alihassan.sodhro@miun.se

***Abstract*—Speech and language relearning are challenging for stroke survivors, as well as medical caregivers. After a stroke, patient's ability to read, write, speak, and listen is decreased to different degrees, which results in a compromised independent life and a decreased quality of life for the patients. Technology-Enhanced Systems (TES) can play a vital role in this context. However, the available software are not explicitly built for recovering stroke patients' needs but often for children's learning needs. This paper is, therefore, aimed at gathering requirements to support the design of speech and language relearning software applications for stroke survivors. A design science approach was adopted, where different stakeholders such as medical caregivers and information technology consultants were involved in the process. Deductive thematic analysis was conducted to analyze the main findings according to adult learning theory. The software requirements vary from patient to patient where the intensity of speech and language impairments, general medical condition, age, prior experience, and knowledge about the previous health record, and social setup of the patient are playing an important role. The speech therapists should have management functionality in the software to adjust the relearning exercises according to the patient's needs. Since, stroke is most common in adults who learn differently than children, andragogy principles were useful in recognizing patients' health conditions. Adults are interested to be involved in the development of their relearning process. Software for speech and language assessment will be helpful to establish relearning objectives for patients.**

Keywords- Requirements specification, E-Health, Stroke rehabilitation, Speech and language relearning, Adult learning theory, Independent living

I. Introduction

Stroke is one of the most common causes of death and different kinds of chronic disabilities in adults [1]. The fundamental reason for stroke is a partial or complete stoppage of blood flow to the brain that severely affects the brain function; consequently, the overall human body may face different types of disabilities [2]. A stroke survivor often suffers from several long-term mental and physical impairments that have a considerable impact on the patient's daily life activities [3]-[6]. Disabilities and their rehabilitation following stroke are generally divided into speech, cognitive, and motoric impairments [7]-[10].

This research paper focuses on speech and language disabilities and the role of TES for relearning speech and language skills. Almost one-third of a stroke survivor's ability past learning experiences, to communicate is reduced due to impaired reading, writing, listening, and speaking skills [11]-[15]. Most impairments occur during the first weeks after the stroke, but the rest of the recovery may take several years. A decreased ability to communicate has some severe consequences for the patient. One of the consequences is an unhealthy social life [16]-[20]. The patients' social life deeply affects their quality of life; they seem to lose the pure happiness of life and often they feel isolated from society [21]-[25]. To reduce the potential risk for a patient's depressed mental state, the process of relearning communication skills needs to be started as soon as possible after a stroke. Several studies highlighted the benefits of early interventions right after the stroke [26]-[30].

Speech and language relearning require a long-term and intensive rehabilitation plan that involves different types of treatments and exercises. These interventions need many human and financial resources in hospitals and rehabilitation centres. The goal of Speech and Language Therapy (SLT) is to improve the patient's speech and language abilities [31]-[34]. There is some evidence that highly intensive, highly dosed, and long-term therapy have better results as compared to low intensity, low dosed and short-term therapy [35]-[39]. The content of the SLT relates to the stroke patient's damage and intention, such as trying to get back to their speech before the stroke or withholding the current situation. One example of SLT is object identification, classified as simple or complex order comprehension. Simple order comprehension can be to "Put the glass close to the plate" and complex "Put the glass close to the plate and the glass close to a fork." However, the supply of required resources seems not to be enough for the drastically increasing number of stroke patients [40]. Therefore, alternative interventions for speech relearning exercises must be explored. Another critical issue with traditional speech interventions is compromised independent living. Generally, stroke survivors need to stay in the rehabilitation centre, and they are heavily dependent on medical caregivers to perform different kinds of exercises. Various studies suggested that living in the home environment has potential benefits for the patients' overall treatment; in fact, the process of rehabilitation and relearning seems to be more efficient and effective in their own homes [41]. In this context, the use of TES can play a vital role to perform speech relearning exercises at home.

Software applications to perform speech exercises may not only decrease the operational costs for medical caregivers, but it may also provide a sense of joyfulness and independence to the patient [42]. However, the acceptance of these applications depends heavily on the degree of trust in TES, eHealth literacy, ease of use of software applications, and patient's integrity. The software applications should be interactive, self-explanatory, and secure so that patients can quickly adopt and trust them [43].

Lack of tailor-made software applications for speech relearning exercises is also an essential factor to consider. Authors' previous studies highlighted that currently used speech relearning applications in rehabilitation centres are actually developed for school-going children rather than adults [4]-[15]. Stroke patients usually face difficulties using these applications, because of differences in context and learning behaviors between adults and children [4]-[16]. The intensity of speech deficiency differs from person to person where some patients may have some minor issues with communication. In contrast, others may not be able to speak even a few words [44]. Therefore, an individualized and tailor-made software application is needed so that it can be easily adapted according to the patients' current physical and mental condition. Stroke is most common in adults; however, commonly used speech relearning applications are not developed from an adult's learning perspective [45]-[47]. The adult learning theory highlights that adults actively participate in the planning, development, and implementation of their learning process [48]. Therefore, adult learning principles should be considered in the requirement identification process.

The study aimed to gather the requirements for designing an interactive speech relearning software application for stroke survivors. The requirements were also considered from the adult learning principles' perspective. The addressed research questions were:

1. What are the requirements for designing an interactive software application for speech relearning exercises following a stroke?

2. How can the principles of adult learning support understanding the patients' needs?

The remaining paper is organized as follows. Section II, addresses the Knowles' adult learning theory model. Andragogy in practice model is presented in Section III. Method is presented in Section IV, while findings and discussion are presented in Sections V and VI respectively. Finally, conclusion and future work are discussed in Section VII.

II. Knowles' Adult Learning Theory

Several studies have successfully used adult learning principles and andragogy (adult learning theory) in the practice model for education, training, and development of adult learners [13]-[19]. Andragogy highlights that adults tend to learn differently than the traditional children's education that is usually referred to as pedagogy [15]. Adults should be involved in the overall process of planning and implementation of learning objectives [20]. Knowles et al. described the following six characteristics of adult learning model that is guidance for them [15].

A. Need to know

Adults need to know the usefulness of learning objectives before they start learning. Adults invest considerable time and energy in exploring the perceived benefits of learning compared to the drawbacks of not learning. Therefore, the first task of the facilitator or instructor should be to bring the need to know to the learner's awareness.

B. Self-concept

Adults are usually self-directed, and they like to take responsibility for their decisions. A person tends to shift his or her self-concept from dependency towards self-direction. Adult learners' active participation and collaboration in the learning process is needed to enhance and stimulate their learning.

C. Learning from experiences

Adults are usually influenced by their past learning experiences, which vary from person to person. The facilitator/instructor should have a good understanding of an adult's previous experiences and beliefs in the given field. Knowles suggests an individualized learning plan for adults according to their previous experiences.

D. Readiness to learn

Knowles emphasizes the importance of task-oriented learning for the social and professional development of adults. The perceived social benefits of a learning task increase its readiness to support learning. The adults feel an urge to learn when they realize their changed circumstances and the role of learning in these situations.

E. Orientation to learning

Adults tend to learn the skills that have a direct impact on their real-life circumstances. Problem-solving tasks and exercises should be involved in the learning process. The focus of learning should be problem-centred rather than subject-centred.

F. Internal motivation

External motivation factors, such as a better job, good grades in education, and a higher salary are essential for learning. However, adult learning is heavily influenced by internal motivation factors, such as increased quality of life, satisfaction and pleasure at work, and self-esteem. Usually, adults are motivated for self-improvement and growth; however, this motivation is often compromised by a lack of resources, time, and violation of adult learning principles.

III. THE ANDRAGOGY IN PRACTICE MODEL

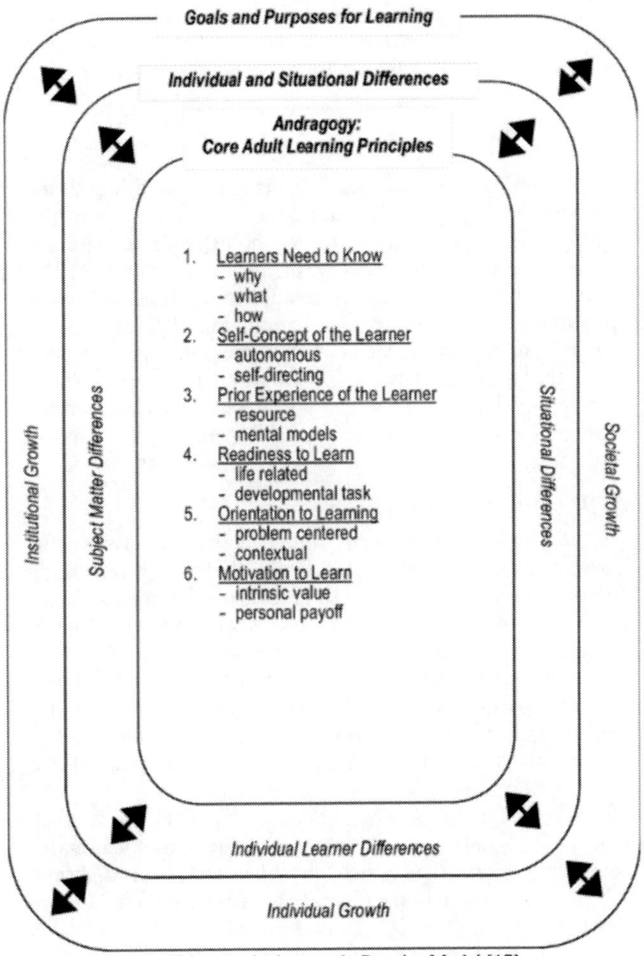

Figure 1. Andragogy in Practice Model [17].

Based on the adult learning principles as described in the previous section, Knowles suggested a conceptual framework that can be adopted for several adult learning practices [15]. Since, most of the stroke patients are adults, adoption of the Andragogy in Practice Model (APM) is a promising approach for speech and language relearning as shown in Figure 1. As shown in the figure, the three dimensions of adult learning in practice may influence the adult learning process.

The outer ring presents goals and purposes for learning that can be seen as developmental outcomes for the learner. The goals can be categorized into individual, social and institutional growth of the learner. The middle ring shows individual and situational differences that might have an impact on learning practices. These differences are further categorized as individual learner differences, subject matter differences, and situational differences. The core six principles of adult learning were used as the primary themes that highlight patients' need for technology-enhanced speech and language relearning. The middle and outer rings of andragogy in practice model were used as filters; the core principles were examined through those filters to the requirements for the technology-enhancement following stroke.

IV. METHOD

The research methodology for this study is a design science research (DSR) that consists of a five-step process described by researchers in [24]. Generally, DSR consists of a rigorous process where a defined problem is solved by designing and implementing an artefact in order to make research contributions [21]-[23]. Since the study is about designing the requirements; a Requirement-Focused Design Science approach was adopted where the first two steps of the process were followed.

A. Data collection

Interviews are the most common approach for data collection and defining the requirements. Semi-structured interviews were conducted with some important stakeholders. The selection of participants is a tedious but essential part of the data collection process. The selection of highly competent and enthusiastic participants plays an essential role in the requirements specification [24]. A purposive sampling approach was adopted for the selection of participants, where all of the participants should have good knowledge and expertise in speech rehabilitation. Interviews were conducted with 11 participants; their professional role and experience in the related field are described below in Table 1. Participants 2-7 were interviewed several times for detailed information and requirements identification.

B. Data analysis

For data analysis, a deductive thematic analysis approach was adopted as suggested by [29]. Interviews, based on audio recordings and transcripts, were carefully explored for coding, and essential features of data that are directly relevant to speech and language relearning were established. The identified codes were examined according to the adult learning theory described in the previous section, and the initial themes were gathered from data. The next step was to select and finalize essential themes. The initial themes were thoroughly reviewed, and the relevant themes that were important to answer the research question were selected. The most relevant and essential themes, such as independent living, tailor-made speech, and language relearning, and technology acceptance was analysed and presented according to Knowles adult learning principles.

TABLE I. STUDY PARTICIPANTS

Participants	Professional role	Years of experience
Participant 1	Speech therapist #1	25
Participant 2	Speech therapist #2	4
Participant 3	Speech therapist #3	5
Participant 4	Stroke specialist doctor and manger in the regional hospital	25
Participant 5	Occupational Therapist	5
Participant 6	Physiotherapist #1	8
Participant 7	Physiotherapist #2	3
Participant 8	Chairman of the local stroke patient organization	3
Participant 9	CEO of a small company working with game-based stroke rehabilitation	25
Participant 10	Hardware and software specialist at a big multinational company	9
Participant 11	Head of Stroke Team	15

C. *Ethical considerations*

Discussing physically and mentally impaired people has always been a sensitive issue when it comes to ethics. Ethical considerations are essential as a researcher, mainly dealing with people in the research and the consequences of the research on those people [26]-[29]. The Swedish Research Council [30]-[35] classified research ethics as professional ethics and categorized them in the following three subcategories: performing a fruitful work, following national and local rules and following the professional codes of ethics [36]-[39]. The third subset describes considerations regarding ethics for collaboration and working environment with co-workers.

At the start of every interview, the interviewees were briefed about the consent of correspondence, including some crucial details about their right to withdraw some specific questions or entire interview at any point. Additionally, they were informed that the purpose of gathering information through interviews is only academic research. The confidentiality of the Participant is also an essential aspect of ethics [28]. The Participants were also briefed that their personal information, such as their names will not be mentioned in the research report. Moreover, the gathered data was safely stored at the university's database, where a strong password is needed for access.

Authors in suggested a close collaboration with the user groups [29]. In order to create a healthy work environment, the researcher should respect the user group, and the users should feel satisfied and secure. Therefore, semi-structured interviews were conducted in a warm environment. Medical caregivers, such as speech therapists were also interviewed for requirement specification. Before conducting the interviews, there were some open discussions between the speech therapists and the researchers to exchange knowledge in their area of expertise. These discussions will help the researchers to create a healthy and secure working environment. The article has been peer-reviewed and discussed at the 43rd Information systems research seminar in [33].

V. FINDINGS

The findings were analyzed thematically according to the Knowles Andragogy in Practice Model (APM); as shown in Figure 1. The patients' needs for technology enhancement are presented using adult learning principles as core themes. The adult learning theory and its principles were described to the speech therapist in regional rehabilitation (Participants 2). Supersizing, they have been considering speech and language from the adult learning aspects without knowing those principles, and they endorsed the idea of involving these principles for technology-enhanced speech and language relearning.

A. *The patient's need to know*

Several informants describe the necessity to involve the patients from the beginning while discussing relearning (Participants 1- 4, 11). They all emphasize the importance of describing the actual situation and what they can achieve. Informant 4 describes this question as to the most common second question from the patients, where the first one is if they are going to survive or not. By setting the goals, the patients can understand what they possibly can achieve by training. The actual situation is described as what happens if the patients skip their training.

The bases for the patient's learning objectives are several. One is the goals and another the patient's motivation for relearning. Highlighted is also the patient's physical condition, which is assessed at a specific meeting with a speech therapist (Participants 1-2). The assessment is conducted as a standardized procedure, involving sets of detailed assignments. Before starting the standardized procedure, the speech therapists decide which parts to assess, based on the patient's described injuries. The assessment relies on an analogue procedure, developed and used in Sweden by speech therapists, and is commonly used throughout the country. Both speech therapists are keen on converting the analogue assessment process into a digital one, where the results would easily be stored and used as input for the relearning assignments.

B. Self-concept of patients

Since adults are usually self-directed, they would like to take responsibility for getting the information about their disabilities and the process of recovery. Several participants highlighted that patients like to get information from their relatives and friends (Participant 2, 5-7). Therefore, the patient's close relatives and friends can play an essential role in the success of a technology-enhanced system. One speech therapist mentioned ongoing research involving the close relatives of patients in the speech relearning process (Participant 2). That research focuses on educating and training the patient's relatives so that they will be able to help patients to perform relearning exercises. The software should have a feature that enables the patient's relatives to collaborate with the patient as well as the speech therapist (Participant 2). An online session with the speech therapist, the patient, and the patient's close relatives such as the husband or wife of the patient can be helpful not only for the patient to perform a different kind of speech relearning exercises but for the speech therapist as well to guide the patient for those exercises.

Personal integrity is also a matter of concern for the stroke survivors while performing therapy from a distance. People do not like to be monitored all the time during the rehabilitation exercises; they want to do the exercises independently as much as possible (Participant 5-7). The patients should be able to use the application independently with the least interaction or guidance from the therapist (Participant 2, 3).

C. Adult patients learn from their experiences

Adults learn from their previous experiences, such as knowledge from previous understandings about TES build a perception of the use of TES. It is, therefore, essential to consider patients' previous practices and knowledge about technology enhancement. Many of the available software for speech is developed for English speakers; however, the speech therapies should be conducted in the native (Swedish) language (Participant 2, 11). The Head of the local mobile stroke team highlighted that the number of immigrants with different languages is increasing in Sweden and healthcare givers need more and more resources for translation services. Therefore, the option of selecting different languages in the software is preferable (Participant 11).

To use speech relearning exercises, education and training are also needed not only for the patients but for the speech therapists as well (Participant 2). Medical staff faced many difficulties in setting up online meetings with medical caregivers; therefore, it might be more challenging for them to guide the patients who are already facing impairments because of stroke (Participant 2, 3). Older people particularly face more problems while using speech relearning exercises on smartphones and tablets (Participant 1). Older adults with limited previous experience of using TES have more difficulties than the younger generation and disability after a stroke makes it even more challenging for them (Participant 1, 4-7).

D. Readiness to learn

After designing the learning strategy and tools, the patients should be prepared for the implementation of the learning strategy. Proper education and training, usability considerations, and social aspects of TES might increase the readiness to learn for patients. Almost all the participants emphasized the importance of technology acceptance requirements such as usefulness, ease of use, adaptability, and satisfaction of the software application (Participants 1-8, 11). Two of the speech therapists informed us that there are some applications available for speech relearning exercises; however, they have not developed specificity according to stroke patient's medical conditions (Participant 1, 2). The tendency to use those applications heavily depends on the degree of impairments after stroke and the overall patient's health (Participant 5-7).

Two speech therapists suggested different levels of login settings for patients and medical caregivers. The software should have a simpler version of the interface for the patients where they can perform their exercises; however, the speech therapists should have a detailed version where they can administer, and suggest and monitor different kinds of speech therapies (Participant 2, 3). The stroke specialist doctor highlighted that the patient should feel a sense of pleasure and satisfaction while using software applications; therefore, goal-oriented training with the element of entertainment can improve the usability of this software (Participant 4). One suggestion, from the chairman of the local stroke organization, is the involvement of music and dance during the relearning exercises (Participant 8).

E. Orientation to learning

The relearning strategy should be individual-focused rather than disability-focused. Therefore, an individualized learning plan is needed for patients with different medical, social, and professional background. Participants with medical backgrounds highlighted that the technical requirements could differ according to the patient's medical condition and their perception about the use of technology enhancement (Participant 1-7). Patients with impaired cognition might have problems using complex text-based interfaces. Therefore, applications with features of recording and replaying can be beneficial for those types of patients, especially for pronunciation training (Participant 1). Patients with severely impaired language skills are recommended to use image-based tools, such as photographs, for communication (Participant 1, 2).

Due to the brain injuries after stroke, the patient's focus for doing different tasks is also decreased. Therefore, exercises with low intensity and a shorter period are more beneficial than high-intensity exercises (Participant 1, 5). In some cases, a patient's vision is blurred after stroke, and they can have a limited view of things; however, the interfaces are mostly designed for the users with full vision (Participant 1, 2).

Therefore, the interface should be designed according to the patient's view level. With severe physical condition, eye-tracking technology might be helpful where patients can navigate the interface through eye movement (Participant 10).

A stroke survivor's choice of hardware also depends upon his physical condition; the software should be usable on different types of devices such as smartphones, tablets, and computers (Participant 1-3). The same requirement is suggested by the hardware and software expert, the application should be platform-independent, and it should be compatible with different kinds of hardware (Participant 10).

F. Internal Motivation of patients to relearn

Participants 1 - 4 emphasized that internal motivation is essential while discussing what can be achieved. For a person involved in managing various situations, such as being a politician or chairman, is the speech of specific interest to continue activities conducted before the stroke as much as possible. Therefore, the patients need to know what they can do to live their lives as much as possible as before the stroke.

Living independently in the home environment has some potential benefits for stroke survivors as it increases their internal motivations for relearning. The mobile stroke team highlighted that most of the patients want to go home as soon as possible and feel secure in the home setup (Participant 5-7). The relearning process is fast in the home environment where patients can get help and inspiration from their loved ones (Participant 2).

VI. Discussion

The primary aim of this paper was to gather the requirements for the development of an interactive speech and language relearning software for stroke survivors. Some essential requirements were gathered with the help of different stakeholders involved in stroke rehabilitation. Especially speech and language therapists (Participant 1-3) played an important role by describing the patients' needs according to their impaired medical condition. The secondary aim of this research was to analyze the requirements of Knowles's adult learning theory perspective. Some other related studies also highlighted the involvement of adult learning principles in the speech and language relearning process might increase the efficiency and effectiveness of relearning for stroke patients [31][32]. Andragogy in Practice Model [15] was used as a theoretical framework for speech, and language relearning was developed. In the following discussion, the main findings from empirical data are discussed from previous literature in the field and adult learning principles viewpoint.

Most of the participants highlighted that the requirements are different for different patients according to their physical and cognitive abilities. Several previous studies on speech and language rehabilitation also highlighted the same factor [6][33][34]. Researchers in [44-45] argued that the intensity of relearning exercises should be according to the patient's focus or concentration level; less intensive and short exercises showed better improvement for the patients with low focus levels6. Authors in [46-48] also highlighted that the individual differences of an adult play an essential role in adult learning, and an individualized learning plan is needed [15]. Therefore, the speech therapist should be able to make an individualized relearning plan for the patients according to their medical condition. The speech therapist should have extra functionalities in software so that they can adjust the relearning exercises according to the patient's needs.

Speech therapists suggest a patient-centered approach; both patients and therapists should be involved in the process of software development. As highlighted by [17], adult learners should be involved in the planning and implementation of their learning objectives. Adults want to understand the learning goals before they start learning and their involvement in defining, planning and implementation of the learning process may increase the effectiveness of learning. The importance of user-central design is already a well-known factor from the literature [5][33]; however, very few studies are conducted where both patients and the therapists are involved in the software design process [4][35]. The requirements should, therefore, be looked at from the user's point of view.

Several participants showed some significant concerns about the usability of software applications. Medical caregivers highlighted that they faced many problems using Technology-Enhanced Systems (TES) such as connectivity errors, audio and video efficiency, and screen sharing issues with an online meeting. Proper education and training of the given software is always an essential aspect of usability [36]. Education and training are not only crucial for patients but the speech therapist as well. Usually, speech therapists recommend and educate patients about the use of rehabilitation applications; therefore, therapists need to get familiar with that application first. From the adult learning viewpoint, education and assistance about learning strategies and tools might increase the readiness to learn; the patients should be prepared for the implementation and use of selected tools and technologies for speech and language relearning.

Another important aspect highlighted by the participants is the involvement of patients' relatives and friends in the relearning process. Patients with impaired physical and mental conditions feel more comfortable, secure, and motivated if their loved ones can be involved in the relearning process [13][18][37]. Usability and usefulness can also be improved by adding social networking features in the application where patients may connect with patients, share their stories and experience, and play online games with each other [38]. The social aspects of learning are also crucial for adult learners, as discussed in [42-43], they argued that the social benefits of a learning task increase its readiness to learn.

Software applications to perform speech exercises may not only decrease the operational costs for medical caregivers, but it may also provide a sense of joyfulness and independence to the patient [2][39-42]. However, the acceptance of these applications depends heavily on the degree of trust in TES, eHealth literacy, ease of use of software applications, and patient's integrity [43-45]. The software applications should be interactive, self-explanatory, and secure so that patients can quickly adopt and trust them [46-48].

VII. Conclusion and Future Work

With the help of some experienced and enthusiastic participants, some essential requirements for technology-enhanced speech and language relearning were gathered. Since stroke is most common in adults and adults learn differently than children, adult learning theory was helpful to understand the patients' needs. Adult patients like to involve in the development of their relearning process. Therefore, a speech and language assessment software should be the first step in technology enhancement where a speech therapist can show the patients the level of their current and/or previous impairments and plan a future strategy for their relearning.

The relearning process should be decided according to patients' internal and external circumstances such as the general medical condition of the patient, intensity of speech and language impairments, and patient's social and professional life. Therefore, speech therapists need extra management functionality in the relearning software so that they can adjust the exercises according to the patient's needs.

This study found essential requirements for the future development of technology-enhanced applications for speech and language relearning tailored for stroke patients. The next planned steps are to design, develop, and evaluate two technology-enhanced applications to support the diagnosis and the relearning process after stroke. Both these applications should be designed and tested with a multi-stakeholder approach involving caregivers, software developers, stroke patients, and stroke patients' friends and family. To carry out the work with a multi-stakeholder approach is essential since a stroke patient's speech relearning journey back to an independent life is a long and tedious one.

Acknowledgment

We are thankful to Mobile Stroke Team, the communal rehabilitation centre and other participants in the study for providing us very fruitful empirical data. In addition, all the authors have equally contributed from Department of Computer and System Science (DSV), Mid Sweden University, Östersund, Sweden

References

[1] T.P., Yamato, J.E., Pompeu, S.M., Pompeu, and L., Hassett, "Virtual reality for stroke rehabilitation", Physical therapy, vol.96, no. 10, pp.1508-1513, 2016

[2] M. Tousignant, J. Macoir, V. Martel-Sauvageau, "Satisfaction with in-home speech telerehabilitation in post-stroke aphasia: an exploratory analysis", J Int Soc Telemed EHealth, vol. 6, no. 1, pp. 1-11, 2018.

[3] S. Palmcrantz., J. Borg, D. Sommerfeld, J. Plantin, A. Wall, Ehn, M., "An interactive distance solution for stroke rehabilitation in the home setting–A feasibility study",. Informatics for Health and Social Care, vol. 42, no. 3, pp.303-320, 2017.

[4] K.Ahlin, A.Ahmad, P. Mozelius, "Determining Testbed Requirements for Technology Enhanced Speech Rehabilitation after Stroke-the Informed Co-workers View Point". In: IARIA GLOBAL HEALTH International Conference on Global Health Challenges. 2019, pp.1-8.

[5] A. Ahmad , K. Ahlin , P. Mozelius., "The Technology-Enhanced Requirements for the Three-Fold Stroke Rehabilitation to Support Independent Living", International Conference on Information and Communication Technologies for Ageing Well and e-Health, Springer, 2019, pp. 142–159.

[6] P. Kesav , S.L. Vrinda, S.Sukumaran., "Effectiveness of speech language therapy either alone or with add-on computer-based language therapy software (Malayalam version) for early post stroke aphasia: A feasibility study", J Neurol Sci, vol. 380, no.2017, pp.137–141, 2017

[7] A.H. Sodhro, Arun Kumar, "An Energy-Efficient Algorithm for Wearable Electrocardiogram Signal Processing in Ubiquitous Healthcare Applications", MDPI Sensors, vol.8, no.3, pp.923, 2018

[8] O.Hege Prag, et al., "Telerehabilitation for aphasia–protocol of a pragmatic, exploratory, pilot randomized controlled trial", Trials, vol. 19, no.1, pp. 208, 2018

[9] M.C.Brady, H. Kelly, J. Godwin, "Speech and language therapy for aphasia following stroke", Cochrane database of systematic reviews, vol.6, no.2016, pp.1-16, 2016.

[10] D. Hu, et al. " Decomposing Atrial Activity Signal by Combining ICA and WABS", 35th IEEE Annual International Conference of the Engineering in Medicine and Biology (EMBC), 2013, pp.1-7

[11] N. Zahid, et al., "An Adaptive Energy Optimization Mechanism for Decentralized Smart Healthcare Applications", 93rd IEEE Vehicular Technology Conference (VTC) 2021-Spring, 25-28 April, Helsinki, Finland, 2021, pp.1-6

[12] A. Ahmad, P. Mozelius , K. Ahlin, "Testbed requirements for technology enhanced stroke rehabilitation to support independent living. Eds'Book Testbed Requir Technol Enhanc Stroke Rehabil Support Indep Living", INSTICC Press 2019 Edn, pp.1-8.

[13] A. Ahmad, P. Mozelius, "On the Importance of Tailor-made Speech Relearning Software for Stroke Rehabilitation", The International Conference on Information and Communication Technologies for Ageing Well and e-Health (ICT4AWE) 2020, 2020, pp.176-179.

[14] S.Eames, et al. "Randomised controlled trial of an education and support package for stroke patients and their carers", BMJ Open, vol.3, no.5, pp. 1-9, 2013.

[15] A.H.Sodhro, Arun Kumar, Gul Hassan Sodhro, "5G-based Transmission Power Control Mechanism in Fog Computing for IoT Devices", MDPI Sustainability, vol.10, no.4, pp.1-17, 2018

[16] O.A.Egaji, et al., " Digital Speech Therapy for the Aphasia Patients: Challenges, Opportunities and Solutions", In: Proceedings of the 9th International Conference on Information Communication and Management. 2019, pp. 85–88.

[17] M.S.Knowles, E.F. Holton III, R.A.Swanson, "The adult learner: The definitive classic in adult education and human resource development", Routledge, 2014, pp.406.

[18] S.B. Chesbro, and L.A.Davis, "Applying Knowles' model of andragogy to individualized osteoporosis education. Journal of Geriatric Physical Therapy, vol. 25, no. 2, pp.1-8, 2002.

[19] A.H. Sodhro, S.Pirbhulal, K. Muhammad, "Towards 6G Architecture for Energy Efficient Communication in IoT-Enabled Smart Automation Systems", IEEE Internet of Things Journal, pp.1-8, 2020 (early access)

[20] E.K. Kaufman, et al., "Leadership development for local volunteers: a case study of andragogy in practice", Leadership, vol.26, no. 3, pp.1-14, 2009.

[21] N.N. Dehghan, et al., "Investigating the effects of a family-centered care program on stroke patients adherence to their therapeutic regimens", Contemp Nurse, 2014;vol. 47, no.2014, pp.88–96, 2014

[22] A.H.Sodhro, et al., "Mobile Edge Computing based QoS Optimization in Medical Healthcare Applications",

International Journal of Information Management , Elsevier, vol.45 , no. 2019, pp.308-318, 2019

[23] M.S.Knowles, et al., "The adult learner: The definitive classic in adult education and human resource development", 6th ed. Burlington, MA, Elsevier, 2005.

[24] P. Johannesson, and E. Perjons, An introduction to design science. Springer, Second Edition, pp.197, 2014.

[25] K. Peffers, T. Tuunanen, M.A. Rothenberger., "A design science research methodology for information systems research", Journal of management information systems, vol. 24, no. 3, pp. 45-77, 2007

[26] A.R.Hevner., S.T.March., J. Park., "Design science in information systems research", MIS Q, vol. 28, no. 1, pp. 75–105, 2004.

[27] A.H.Sodhro, et al., "Artificial Intelligence based QoS optimization for multimedia communication in IoV systems", Future Generation Computer Systems, Elsevier, vol. 95, no.2019, pp. 667-680, 2019

[28] M. Muzammal, et al., "A Multi-sensor Data Fusion Enabled Ensemble Approach for Medical Data from Body Sensor Networks", Information Fusion, Elsevier, vol.53, no.2020, pp.155-164, 2020

[29] A.H.Sodhro, Y.Li, M.A.Shah, "Energy-efficient Adaptive Transmission Power Control in Wireless Body Area Networks", IET Communications, vol.10, no.1, pp.81-90, 2016

[30] S.Kvale, et al., "The qualitative research interview, pp. 412, Third Edition, Lund: Studentlitteratur, 2014

[31] A. H. Sodhro, A. K. Sangaiah, S.Pirphulal, "Power Management Strategies for Medical Information Transmission in Wireless Body Sensor Networks", IEEE Consumer Electronics Magzine, vol.9, no.2, pp.47-51, 2020

[32] A.Ahmad, K. Ahlin, P. Mozelius, "Technology-enhanced Speech and Language relearning for stroke patients-defining requirements for a software application development", 11th Scandinavian Conference on Information Systems (SCIS2020), Sundsvall, Sweden, 2020, pp.1-7.

[33] M.L.Kimbarow, "Integrating life participation approaches to aphasia treatment with adult learning theory: A synergistic approach", Top Lang Disord, vol. 27, no.2007, pp. 318–323, 2007.

[34] T. Hopper, A.L. Holland, "Aphasia and learning in adults: Key concepts and clinical considerations", Top Geriatr Rehabil, vol. 21, no. 2005, pp.315–322, 2005.

[35] M.Muzammal, R.Talat, 'A Multi-sensor Data Fusion Enabled Ensemble Approach for Medical Data from Body Sensor Networks', Information Fusion, Elsevier, Vol. 53, No.2020, pp.155-164, 2020

[36] T.Simic T, C. Leonard, L. Laird et al. A usability study of internet-based therapy for naming deficits in aphasia. Am J Speech Lang Pathol 2016; 25: 642–653.

[37] Y.Lin, et al., "An analytic computation-driven algorithm for Decentralized Multicore Systems", Future Generation Computer Systems, vol.96, no.2019, pp.101-110,2019.

[38] A.D. Dabbs, et al., "User-centered design and interactive health technologies for patients". Computers, informatics, nursing: CIN, vol. 27, no. 3, pp.175, 2009

[39] K. Mallet, R. Shamloul, M. Pugliese., "A patient perspective on the delivery of mobile Tablet-based stroke rehabilitation in the acute care setting". Int J Stroke, vol. 14, no. 2019, pp. 174–179, 2019

[40] S. Mhammadi, et al., "The Effect of Family-Oriented Discharge Program on the Level of Preparedness for Care-Giving and Stress Experienced by the Family of Stroke Survivors", Iran Rehabil Journal, vol. 17, no.1, pp.113–120, 2019.

[41] M.B.Garcia, "A Speech Therapy Game Application for Aphasia Patient Neurorehabilitation–A Pilot Study of an mHealth App", International Journal of Simulation: Systems, Science & Technology, vol.20, no.2, pp. 1-9, 2019

[42] D. Veer, et all., "Determinants of the intention to use e-Health by community dwelling older people", BMC health services research, vol.15, no. 1, pp.103, 2015.

[43] A.H. Sodhro, et al. "Decentralized Energy Efficient Model for Data Transmission in IoT-based Healthcare System" , 93rd IEEE Vehicular Technology Conference (VTC) 2021-Spring, 25-28 April, Helsinki, Finland, 2021, pp.1-5

[44] A. Ahmad, P. Mozelius, 'Critical factors for human computer interaction of ehealth for older adults', Proceedings of the 2019 the 5th International Conference on e-Society, e-Learning and e-Technologies. 2019, pp. 58–62.

[45] L.T.Vassli, B.A Farshchian, "Acceptance of health-related ICT among elderly people living in the community: A systematic review of qualitative evidence". Int J Human–Computer Interact, vol. 34, no.2018, pp. 99–116, 2018.

[46] H.Magsi, et al., "A Novel Adaptive Battery-Aware Algorithm for Data Transmission in IoT-Based Healthcare Applications", Electronics, vol.10, no. 4,pp.367, 2021

[47] I. Ahmad, et al., 'Machine Learning Meets Communication Networks: Current Trends and Future Challenges', IEEE Access, vol.8, no. 2020, pp. 223418-223460, 2020

[48] A. H.Sodhro, et al., "Medical-QoS Telemedicine Service Selection using Analytic Hierarchy Process", Handbook on Smart Healthcare, Springer , Handbook of Large-Scale Distributed Computing in Smart Healthcare pp 589-609, 2017

Public Awareness and Acceptance of Telemedicine in Japan

Nermin Elokla
Faculty of Applied Arts
Helwan University
Cairo, Egypt
email: nelokla@hotmail.com

Tomohiko Moriyama
International Medical Department (iMed)
Kyushu University Hospital
Fukuoka, Japan
email: tomohiko.moriyama.153@m.kyushu-u.ac.jp

Naoki Nakashima
Medical Information Center (MIC)
Kyushu University Hospital
Fukuoka, Japan
email: nnaoki@info.med.kyushu-u.ac.jp

Abstract—Telemedicine seeks to improve the quality, efficiency and cost of healthcare by a variety of electronic means. It is originally designed to serve patients in remote areas. The service involves the use of technologies to support the connection between doctors and their patients without the need for in- person meetings. Nowadays, the COVID-19 outbreak could put a spotlight on Telemedicine as an important tool to avoid the community from infection. This study aims to examine public awareness and acceptability of telemedicine in Japan. Furthermore, it determines the main reasons which hinder the growth of the service in the COVID-19 era from the public's views. To achieve our goals, 84 participants in the age ranging from 20 to 64 years were randomly selected, and data were collected from them using a questionnaire survey. The survey reveals that all participants are not telemedicine users. Fifty-nine (70%) participants are somewhat aware of telemedicine service, and 51% of them are very much interested in telemedicine. Only 3 (4%) participants are quite satisfied with the amount of information available on telemedicine. This is obvious as 50% of them have no detailed information about the service. Fifty-two (37%) participants state that online telemedicine education program is an effective way to raise awareness of telemedicine. The survey results also identify the major reasons for not using telemedicine service in the COVID-19 era. At present, there are various issues for a slow growth of telemedicine in Japan. Recommendations to increase awareness of telemedicine service have been made.

Keywords-telemedicine; service development; Japanese people; awareness and perceptions; COVID-19 era

I. INTRODUCTION

Telemedicine is defined as the use of Information and Communication Technologies (ICTs) to improve patient outcomes by providing accessible, cost-effective, and high-quality healthcare services [1]. Telemedicine uses ICTs mainly to liberate patients and medical professionals from geographical barriers and time restrictions. The recent concept of telemedicine is extremely wide, ranging from diagnosis to therapeutics [2][3]. With the world-wide dissemination of telemedicine, as well as its clinical benefits and cost effectiveness, the Japanese Ministry of Health, Labor and Welfare (MHLW) has made efforts to promote telemedicine since 2015 [4]. In 2018, the service became available across the country but was limited to patients with certain chronic conditions who had already received prescriptions for medication [5]. Studies have been published showing the effectiveness of telemedicine in the treatment of various chronic diseases, such as hypertension, dyslipidemia, and diabetes [4]. In April 2020, the COVID-19 crisis has prompted Japan to ease regulations on telemedicine [6][7]. However, the use of deregulated telemedicine has been sluggish compared to the US and UK [8]. Therefore, the present study aimed to examine public awareness and acceptability of telemedicine. Furthermore, it determined the major reasons which hinder the growth of telemedicine from Japanese people's views. The rest of this paper is organized as follows. Section II explains the method of this study. Section III shows the major results of the survey. Section IV includes discussion and limitation of a study. Section V includes both conclusion and future work.

II. METHOD

To carry out this study, semi-structured questionnaire (12 multiple-choice questions) was undertaken with 84 participants (63 Female and 21 male) between February and April 2021. Selecting the survey participants was based on their busy works with limited vacations time (about 10 days in a year) to visit hospitals. Participants were provided with an explanation of telemedicine and all information regarding the study, including the reasons for undertaking the survey. The questionnaire sheets had been given to the person of charge by hand in order to distributed among the employees. The questionnaire guide was informed by a literature survey in this study area [9][10] and piloted on 5 individuals. Following this process, questions were revised and determined. Probing questions were used to explore public awareness, and acceptability of telemedicine. Furthermore, the questions were used to find out the main reasons that hinder the use and growth of telemedicine in the COVID-19 era. Ethical approval for this study was obtained from the Kyushu University Hospital, Permission No 2021-15.

III. RESULT

A total of 84 administrative employees of which female (63) and male (21) completed the questionnaire. The following are the main results of the questionnaire.

A. *The Characteristic of Participants*

The participants' ages are ranging from 20 to 63 years. They are administrative employees working (full-time job)

TABLE I. CHARACTERISTICS OF PARTICIPANTS (N=84)

Gender N=84	Age	N=84	Physical Disability	City	Occupation	N=84	Hospital visit	(in a year)	N=84	Experience of Telemedicine
Female N 63 (75%)	20-29	17 (20%)	No	Fukuoka	Administrative employees		Once a week	52 times	2 (2%)	No
	30-39	16 (19%)			University A	N 53 (63%)	Once every 2 weeks	26 times	2 (3%)	
	40-49	28 (34%)			University B	N 31 (37%)	Once a month	12 times	16 (19%)	
	50-59	17 (20%)					Once every 2 months	6 times	6 (7%)	
Male N 21 (25%)	60-69	6 (7%)					Once every 6 months	2 times	24 (29%)	
							Once a year	One time	12 (14%)	
							No visit	0	0	
							No answer	—	22 (26%)	
100%	100%				100%				100%	

in different business sectors at 2 public universities in the Fukuoka city. The survey results indicated that all participants have no physical disabilities, and they had never experienced telemedicine. The majority (29%) visit the hospitals about 2 times a year (Table I).

B. Awareness of Telemedicine

About the levels of participants' awareness of the service, 59 (70%) participants were somewhat aware of telemedicine (Table II). Participants' awareness of telemedicine developed through various means. The majority (53%) of participants indicated that media is the main source to know about telemedicine. Only 3 (4%) participants were quite satisfied with the amount of information available on telemedicine. This is obvious as 43 (50%) participants had no detailed information about the service. Furthermore, 68 (56%) participants requested to clearly know about the regulations, cost, benefits, and risks of telemedicine service.

Concerning the acceptability of telemedicine, about 42 (51%) respondents were quite interested in telemedicine concept, technologies and communication mode (Table III). Seventy-two out of eighty-four participants believed that telemedicine has various benefits. In Figure 1, the majority (30%) of participants indicated that it is an effective tool to avoid a risk of hospital-acquired infections.

Regarding the possibilities to raise public awareness of telemedicine, 52 (37%) participants called for setting up online education programs to teach people about telemedicine service (Table II).

TABLE II. PARTICIPANTS' AWARENESS OF TELEMEDICINE (N=84)

Issues	Responses (N=84)				
	To a great extent	Somewhat	Not at all		
Current perception of telemedicine	14 (17%)	59 (70%)	11 (13%)		
People satisfaction with the amount of information available on telemedicine	3 (4%)	59 (70%)	22 (26%)		
Information sources	Media	Hospital	Website	No answer	Friends
	51 (53%)	20 (21%)	12 (12%)	11 (12%)	2 (2%)
Available information	Positive perception	Negative perception	No detailed information	Other	
	31 (36%)	6 (7%)	43 (50%)	6 (7%)	
Additional required information	Telemedicine regulations, benefits, risks, cost, etc.	Patients' experiences and satisfactions with telemedicine	How telemedicine is worked; such as tools, types, etc.	Other	
	68 (56%)	36 (29%)	16 (13%)	2 (2%)	
Possibilities to raise the awareness of telemedicine	Set up telemedicine education program for public	Encourage the use of eHealth through social media, website, email, brochures, etc.	Make the use of telemedicine mandatory by policy or other means	Other	
	52 (37%)	51 (36%)	29 (21%)	8 (6%)	

TABLE III. ACCEPTABILITY OF TELEMEDICINE (N=84)

Issues	Responses		
	Very much	Not much	Not at all
Accepting to the concept of telemedicine	48	24	12
Impressed with telemedicine technology	37	41	6
Interested in virtual/online communication	41	38	5
Average	42 (51%)	34 (41%)	7 (8%)

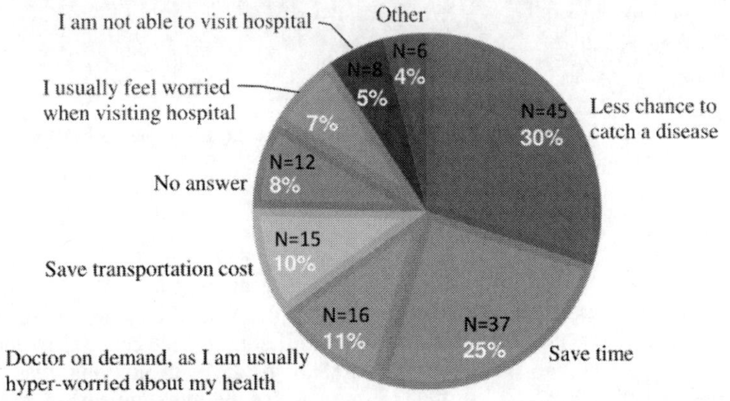

Figure.1 Benefits of Telemedicine from the Japanese People's Perspectives (N=72) - Multiple-choice Question

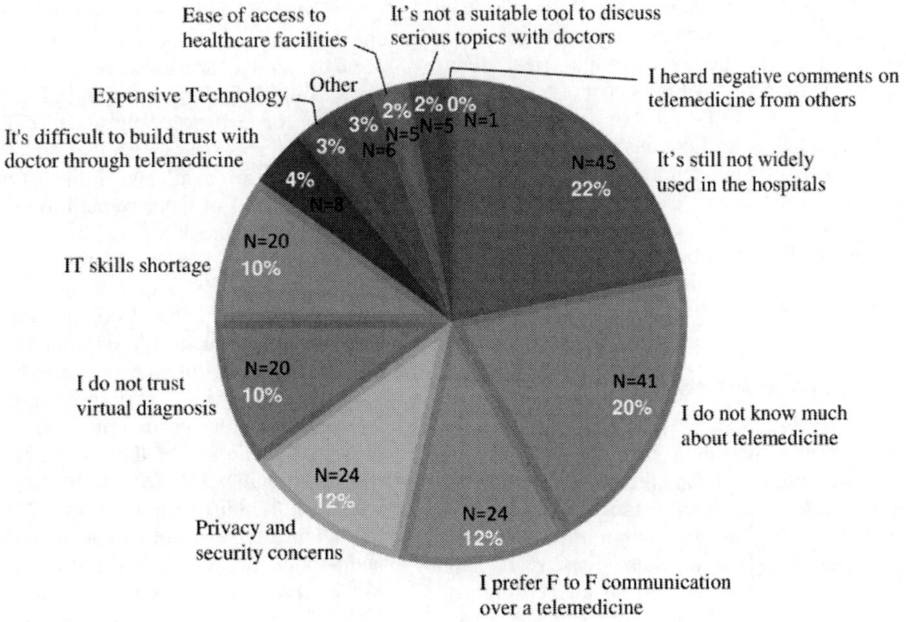

Figure. 2 Why Japanese People Aren't Using Telemedicine at Present (N=84) - Multiple-choice Question

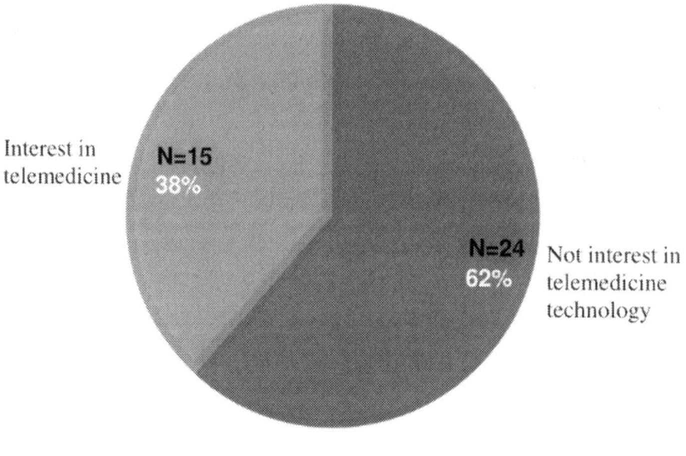

Figure. 3 Participants' Ages (from 45 to 63 years) and Acceptability of Telemedicine (N=39)

C. *Major Reasons for Not Using Telemedicine in the COVID Era*

This study detected that there are different reasons (see Figure 2) led participants to not use telemedicine services. The majority (22%) of participants indicated they do not use telemedicine because it is not widely provided in the hospitals. The study findings show that 20% of participants do not know much about telemedicine services. About 12% of participants preferred to use in-person visit over a telemedicine, while some (12%) participants have concerns about their data privacy in telemedicine. About 10% of participants do not trust virtual diagnosis, whereas other (10%) participants indicated their ICT skills shortage. Furthermore, 4% of survey participants responded that it is difficult to build trust with doctors through telemedicine, while 3% of them mentioned about high cost of telemedicine technology.

On the other hand, 3% of participants chose 'Other'. A few (2%) participants mentioned about ease of access to healthcare facilities, while other (2%) participants indicated that telemedicine is not a suitable tool to discuss serious topics with doctors. Only one participant pointed out that she did not use the service because of receiving negative comments on the service from others.

IV. Discussion

The survey results revealed that the majority of participants are interested in telemedicine. They indicated that telemedicine might be an effective tool to reduce the risk of infection. However, they are not using it at present due to some important reasons. One of these reasons is about providing telemedicine services in a few hospitals. In Japan, there is a slow spread of telemedicine in the hospitals partly due to reimbursement challenges, lack of economic advantage, and a lack of clinical evidence [4][11][12]. According to the MHLW, only 65 telemedicine-based episodes of medical care are performed per month in Japan [4]. Furthermore, a recent study reported that among the 110,898 medical institutions that exist, the number of medical institutions implementing telemedicine increased slightly from 10,812 (9.7%) in April 2020 to 16,202 (14.6%) in June. Of this number, only 6,801 (6.1%) medical institutions implement telemedicine for a patient's first visit [13]. Second major reason is related to limited information available about telemedicine service. Regarding this issue, the survey results detected that a few (4%) participants were quite satisfied with the amount of information avaliable on the service. This is obvious, as the majority (50%) did not have detailed information about telemedicine regulations and risks. Similar findings showed in our pervious study that the majority of Japanese physicians (39%) had no detailed knowledge about telemedicine, and only 3% of them quite satisfied with available information. Furthermore, the majority (32%) of them wanted to know about telemedicine guidelines and standards [12].

Third reason is concerning communication preferences. Although a recent study demonstrated that patients can achieve the same level of communication effectiveness with their physicians using IT communication as they would in comparable Face-to-Face communications [14], about 12% of our participants preferred in-person meetings over telemedicine for promoting conversation through body language, while 4% of them indicated the importance of F-to-F communication for establishing trust in relationship with doctors. Furthermore, a few (2%) participants believed that telemedicine is not an adequate tool for discussing deep and serious topics. Fourth reason is about lack of confidence in the service among the participants. The study revealed that 12% of participants had concerns about protection of their medical information collected by telemedicine. Furthermore,

10% of them did not trust the accuracy of diagnosis by telemedicine. Similar finding showed in other articles that healthcare professionals distrust a diagnosis made via telemedicine [12][15]. Regarding this issue, if patients don't trust the diagnoses made during telemedicine calls, they may ignore the advice given, fail to take preventative steps, or seek additional in-person appointments, which defeats the point of telemedicine [16]. For many e-services, end-user trust is a crucial prerequisite for use [17]. The fifth reason is related to the ICT skills shortage. About 10% of the participants did not have a sufficient technical skills to use telemedicine services. Regarding the acceptability of telemedicine technology, the present study found that the majority (62%) of older participants with ages ranging from 45 to 63 years were not somewhat interested in the telemedicine technology (see Figure 3). Similar finding showed in other study that the elderly had a difficult time adjusting to the new technology of telemedicine [18]. Regarding the last reason, a few (2%) participants indicated that they did not use the service due to the ease of access to healthcare facilities and getting high-quality care at a low fee. Actually, Japan's health system is well known for achieving one of the world's highest life expectancy with universal health coverage. The important feature of the healthcare system in Japan is "free access." Patients are completely free to choose any healthcare facilities, regardless of the severity of their disease and their insurance status [19]. Due to COVID-19 pandemic, a recent survey by the Association of Japan Medical Colleges revealed that the number of patients fell during April and May 2020 because many people opted not to visit the hospital due to fears they might become infected. The number of outpatients in April 2020 dropped by 21 percent over April 2019, while the decrease in May 2020 was 27 percent [20].

Based on the above-mentioned findings, it can be said that a slow growth of telemedicine in the COVID-19 era is due to many issues. Comparing to the results of other studies [11][15], the present study revealed that the major reasons are not only link to the lack of infrastructure or reimbursement challenges but also due to other various reasons, and the most important of which is related to a lack of detailed information provided for all (including ordinary people and clinicians) about the service.

For people to ask for telemedicine service, they need to understand well its possibilities and limitations. According to the participants' opinions, the most effective way to promote the level of awareness of telemedicine is through education programs. Although international remote training programs for medical engineers have been implemented, there are no evidence reports on conducting online education programs for the public in Japan [21]. Therefore, this study recommends establishing telemedicine education program to teach people about telemedicine as an option for 'a quick way' to receive care. One of the necessary roles of the program is providing individuals with appropriate and complete information on when, and how to use telemedicine, as well as legal and ethical aspects associated with the service delivery. This would probably change their opinions about the service.

Regarding the limitation of this study, the survey was done with limited number of Japanese participants, and the results cannot be generalized beyond the participants of a study. The participants expressed their own perspectives for raising public awareness of telemedicine. These may not express views of the majority of Japanese people.

V. CONCLUSION

The present study examined public awareness and acceptability of telemedicine service in Japan. Furthermore, it determined the main reasons which hinder the growth of telemedicine in the COVID-19 era. Based on the survey results, it can be said that limited awareness of telemedicine plays an important role in a slow growth of the service. Greater public awareness about the service's regulations, benefits, and risks is an essential step for improving telemedicine market in Japan. Therefore, this study recommended setting up education programs to provide people with detailed information about telemedicine service. Further researches are recommended to discuss the development of telemedicine education and training programs for all, including ordinary people and clinicians.

In the future, the authors would seek to run a large-scale survey in order to pick up major and minor barriers to telemedicine development in Japan, considering details about the chronic diseases of the population sample.

REFERENCES

[1] World Health Organization, "Telemedicine: Opportunities and developments in member states: Report on the second global survey on eHealth". Global Observatory for eHealth Series, vol 2, 2010. [Online] www.who.int/goe/publications/goe_telemedicine_2010.pdf

[2] T. Reed, E. Margo, and H. Michael, "Telehealth", The New England Journal of Medicine, 377, pp.1585–1592, 2017.

[3] G. Foldgern, A. Rachas, A. Farmer, M. Inzitari, and S. Shepperd, "Interactive Telemedicine: Effects on Professional Practice and Health Care Outcomes", Cochrane Database Syst Re, CD002098, 9, 2015.

[4] Y. Kadoya, M. Hara, K. Takahari, Y. Ishida, and M. Tamaki "Disease Control Status and Safety of Telemedicine in Patients with Lifestyle Diseases" - A Multicenter Prospective Observational Study in Japan - Control Disease Status and Safety of Telemedicine in Patients - with Lifestyle Diseases", Health Services and Outcomes Research, 2, pp. 351–356, 2020.

[5] S. Xincheng, "Multilingual Telehealth Services Target Foreigners in Japan", KYODO NEWS, March 2021.

[6] With APPS and Remote Medicine, "Japan Offers Glimpse of Doctor Visits in Post-Corona Era", The Asahi Shimbun, 2020.

[7] Vial Signs: Digital Health Law Update, April 2020. [Online] www.jonesday.com/en/insights/2020/04/vital-signs-digital health-law-update-spring-2020

[8] Nihon Keizai Shimbun, HEAD TOPICS, August 2020. [Online] https://headtopics.com/jp/369603854835386534728-14820914

[9] N. Bradford, L. Caffery, and A. Smith, "Awareness, Experiences and Perceptions of Telehealth in a Rural Queensland Community" BMC Health Serv Res, vol 15, 427, 2015.
[10] A. Albarrak, et al. "Assessment of Physician's Knowledge, Perception and Willingness of Telemedicine", Journal of Infection and Public Health, vol 14, Issue 1, pp 97-102, 2021. [Online] https://www.sciencedirect.com/science/article/pii/S187603411930139X
[11] P. Branch, S. Wilson, and L. Mills, "Japan 2019- 2020 Hospital Insights Survey", Special Report, 2020. [Online] https://www.lek.com/sites/default/files/insights/pdf-attachments/Japan-Hospital-Survey.pdf
[12] N. Elokla, T. Moriyama, and N. Nakashima, "Why Japanese Physicians Are Not Using Telemedicine in the COVID-19 Era?", The 25th ISfTeH International Conference, Japan, pp. 58, February 2021.
[13] Y. Matsuyama, "Aging, Safety Net and Fiscal Crisis in Japan. no.270: Telemedicine Has not Become More Popular, Despite the Spread of COVID-19", the Canon Institute for global studies, September 2020. [Online] https://cigs.canon/en/article/20200901_5322.html
[14] T. Mirzaei and N. Kashian, "Revisiting Effective Communication Between Patients and Physicians: Cross-Sectional Questionnaire Study Comparing Text-Based Electronic Versus Face-to-Face Communication", Journal of Medical Internet Research, vol 22, 5, 2020.
[15] US JAPAN Business Council, "Leveraging Digital Technologies to Improve Health Care Outcomes: An Opportunity for Policy Collaboration in the U.S., Japan Bilateral Relationship", 2020.
[16] HITC, "75% of Patients Wouldn't Trust a Diagnosis via Telemedicine", 2015.
[17] L. Velsen, M. Tabak, and H. Hermens, "Measuring patient trust in telemedicine services: Development of a Survey Instrument and its Validation for an Anticoagulation Web-Service", International Journal of Medical Informatics, vol 97, pp. 52-58, 2017.
[18] Y. Tamura, et al. "How COVID-19 Affected the Introduction of Telemedicine and Patient Reported Outcomes Among Patients with Pulmonary Hypertension", A Report from a Referral Center in Japan, vol 2, 9, pp. 526-530, 2020.
[19] D. Kato, et al. "Building Primary Care in Japan: Literature Review", Journal of General and Family Medicine, vol 20, 5, 2019.
[20] N. Himeno, "Hospitals Rack Up Massive Losses Trying to Cope With COVID-19", the Asahi Shimbun, July 2020. [Online] www.asahi.com/ajw/articles/13564773
[21] K. Kudo, T. Moriyama, S. Tomimatsu, S. Ueda, and S. Shimizu, "Train to Trainers (TtT) Program for Engineers for Hosting Multiparty International Clinical Teleconference", Journal of the International Society for Telemedicine and eHealth, vol 7, 2019.

IoT Platform for Ageing Society: the SMART BEAR Project

Smart Big Data Platform to Offer Evidence-based Personalised Support for Healthy and Independent Living at Home

Alessia Cristiano, Sara De Silvestri, Stela Musteata, Alberto Sanna, Diana Trojaniello
Center for Advanced Technology in Health and Wellbeing
IRCCS Ospedale San Raffaele
Milan, Italy
e-mail: trojaniello.diana@hsr.com

Valerio Bellandi, Paolo Ceravolo
Computer Science Department
Università degli Studi di Milano
Milan, Italy
e-mail: ceravolo.paolo@unimi.it

Matteo Cesari
Istituti Clinici Scientifici Maugeri
Università degli Studi di Milano
Milan, Italy
e-mail: macesari@gmail.com

Abstract — **People over 65 years old represent a constantly increasing age-group in Europe and worldwide. Hearing loss, cardio-vascular diseases, cognitive impairments, mental health issues and balance disorders are the most prevalent health challenges experienced by older population. These conditions result in a quality-of-life worsening (e.g., inactive lifestyle, missed self-sufficiency, loneliness) and life threats (e.g., physical injury, disability, and hospitalization). Moreover, their management is burdensome for healthcare institutions with high and rising costs, and gaps in quality, safety and access. Internet of Things (IoT) technology may offer a valid aid with its innovative and connected solutions. In such context, SMART BEAR Horizon 2020 European project aims to design and develop an IoT platform to: (*i*) integrate off-the-shelf smart consumer and medical devices for a smart health environment, (*ii*) provide an affordable, secure, and privacy-preserving service to older subjects promoting autonomy and healthy living, and (*iii*) increase the efficiency of healthcare delivery reducing resource waste. The present study is intended to comprehensively illustrate the project. In particular, it summarizes rationale and features of SMART BEAR project with a finer description of the platform components, clinical scenarios, and interventions.**

Keywords - ageing population; IoT platform; personalized interventions; independent living; healthy lifestyle.

I. Introduction

Ageing population represents a major challenge in nowadays societies. According to [1], 151 million of the European population will be over 65 years old by 2060 with a particularly rapid increase in number of people over 80 years old. The physiological and progressive decline due to the ageing in physical abilities (e.g., osteoporosis, frailty) and cognitive skills (e.g., memory and visual-spatial difficulties) leads to a reduced independency and a need of care. Such conditions, in turn, usually cause a deterioration of individual's mood and social participation (e.g., older people are at higher risk of depression and poor self-rated health when compared with younger subjects). As a result, multiple comorbidities of chronic and neurodegenerative diseases are simultaneously faced by older population with heavy repercussions on health policies and programs as well. Actions across multiple sectors enabling older people to remain a resource for their families, communities and economies are therefore needed. The World Health Organization (WHO) has lately set as a priority the promotion of "ageing well" [2] intended as the process for fostering and maintaining the individual's functional ability that includes managing their own basic needs, making decisions, being active, building and maintaining a social life, and contributing to the society [3].

Smart Healthcare (SH) systems addressed to older adults represent valuable solutions as they monitor and track individual's behavior and health status. They can also help in managing medical conditions effectively with a consequently reduced burden on caregivers, and lastly, they can motivate subject to stay physically and cognitively fit so to enhance wellbeing and increase Quality of Life (QoL).

Thanks to the recent advancements of technology, SH systems are increasingly proposed. From 2016 onward [4], a significant increase in the development of smart platforms tailored on ageing population with applications devoted to neurology, cardiology, psychiatry, and psychology has been observed. Those platforms [5] include Internet of Things (IoT) devices that can be worn by subjects (e.g., wearable sensors and medical devices) and/or installed in house settings (e.g., smart home devices). The data collected by the devices are then sent to a cloud system and here analyzed through sophisticated Machine Learning (ML) and Artificial Intelligence (AI) algorithms to properly assess the status of older individual and provide him/her feedbacks or real-time alerts. Although those infrastructures may appear simple, they hide tough tricks because of [5]: *a*) wide heterogeneity of equipment (i.e., wearable sensors, home sensors and medical devices), *b*) big volumes of data usually acquired in uncontrolled environments, *c*) multifactorial selection of the most suitable algorithm for the specific topic to figure out and *d*) privacy and security issues related to the processing of subjects' data.

Within such context, SMART BEAR project, funded by the European Commission under the Horizon 2020 program, aims to design, and develop an innovative platform (i.e., SB platform), integrating state-of-the-art devices, addressed to older adults [6] to offer evidence-based support for healthy and independent living at home. The project involves 27

European partners among universities, research centers, hospitals, collective structures, local administrations, small and medium-sized enterprises and big companies and it will be concluded in 2024 after 60 months of synergistic work.

The aim of the present paper is to provide an exhaustive bird's-eye view on the SMART BEAR project. The document is organized as follows. Section II reports rationale and goals of SB platform. Section III addresses project pilot sites. Section IV illustrates the devices comprising the platform and the principal features of the cloud system. Interventions and related clinical scenarios are described in Section V while conclusions are reported in Section VI.

II. RATIONALE

Within the European ageing population, Hearing Loss (HL), Cardio-Vascular Diseases (CVDs), Cognitive Impairments (CIs), Mental Health Issues (MHIs), and Balance Disorders (BDs) are prevalent conditions [7].

HL [8] is the third most common condition affecting older adults and the fifth leading cause of disability worldwide. It increases the risk of cognitive decline, mental illness, and depression, and it leads to social isolation.

CVDs [9] are the main causes of death globally. They have a physical, social, and emotional impact on older adults and a significant economic impact on European economy as reported by the European Society of Cardiology (ESC).

CIs [10] are prevalent in adulthood, especially after heart failures. They affect several domains as memory, attention, executive functions, and psychomotor speed thus compromising the ability to think, learn and remember.

MHIs [11] include social isolation, sleep disturbances, anxiety, and depression. They have an impact on physical health and vice-versa. In addition, they increase the risk of converting mild cognitive impairments to dementia.

BDs [12] are consequences of age-related progressive loss of sensory information functioning and body movement control. They result in falls, disability, and death (one older adult deaths because of a fall every 29 minutes).

The above-mentioned conditions have hence huge impact on older subjects' QoL and healthcare institutions' finances. Preventing and/or slowing the development of those impairments is therefore beneficial for older people, caregivers, clinicians, and hospitals. Therefore, the main objective of SB platform is to integrate heterogeneous sensors and assistive medical devices that will enable continuous and non-continuous monitoring in older adults' everyday life. The monitoring will be done in both indoor and outdoor environments, depending on every device's feature, and will help to gain evidence needed to plan personalized interventions promoting an independent and healthy lifestyle. The secondary objective, instead, is to identify subgroups' patterns across different countries and urban areas with different extensions, local services, environmental, and socio-economic conditions.

III. PILOT SITES

Throughout the project, the SB platform will be employed and assessed in five large-scale pilots (i.e., Greece, Italy-Portugal, France, Spain, and Romania) with the overall involvement of 5100 older adults living at home and in collective structures (Table I). The high sample size considered enables a large-scale validation of the platform while the multiple geographic areas selected offer a great heterogeneity of conditions allowing comparisons suitable for patterns detection among and within regions.

TABLE I. SMART BEAR PILOTS

Pilot Sites	Sample Population	Sample Size	Geographic Area
Greece	Independent older adults living at home/ Older adults living in collective structures	1000	Region of Peloponnese, Municipality of Palaio faliro
Italy-Portugal	Independent older adults living at home/ Older adults living in collective structures	1100	Milan metropolitan city, District of Crema, Madeira Island
France	Independent older adults living at home/ Older adults living in collective structures	1000	Ile-de-France (the Paris Region), Nouvelle Aquitaine and Bretagne
Spain	Independent older adults living at home/ Older adults living in collective structures	1000	Barcelona, Madrid, Sevilla areas, Pais Vasco, Galicia and Balearic Island
Romania	Independent older adults living at home/ Older adults living in collective structures	1000	Bucharest, Cluj Napoca and Constanta metropolitan areas

IV. SMART BEAR DEVICES AND CLOUD SYSTEM

SB platform integrates sensors and devices able to record data from the daily living of its older users. The technological devices of platform can be divided into:

- personal devices (Figure 1);
- smart home devices (Figure 2).

All of them communicate through Bluetooth and/or Wi-Fi with a dedicated mobile application (i.e., SB mApp) installed on a Samsung Galaxy S10 smartphone and available in iOS environment, too.

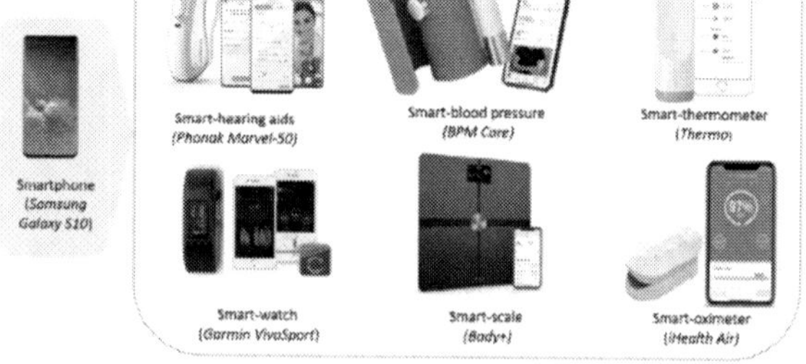

Figure 1. Personal Devices

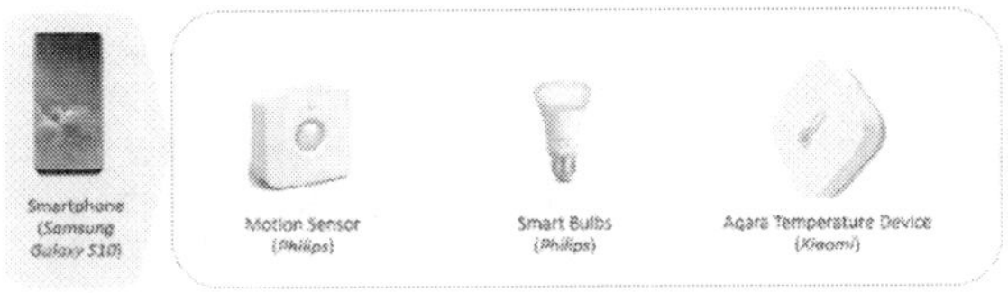

Figure 2. Smart Home Devices

All the personal devices shown in Figure 1 includes:

- ***Samsung Galaxy S10***: smartphone (Samsung Electronics Ltd., South Corea) with a RAM of 3GB up to 1000 GB with a micro SD.
- ***Phonak Marvel-50***: smart-hearing aids (Phonak, Switzerland) rechargeable via batteries with a battery life up to 5-7 days depending on usage.
- ***BPM Core***: smart-blood pressure monitor (Withings, France; dimensions: 560x165x450mm; weight: 430g) rechargeable via micro-USB cable with a duration of approximatively 6 months.
- ***Thermo***: smart-thermometer (Withings, France; length: 116mm, diameter: 66.2mm; weight: 75g) rechargeable via AAA batteries with a battery life up to 2 years.
- ***Garmin VivoSport***: smartwatch (Garmin, USA; width: 21mm, thickness: 10.9mm; weight: 24.1g) with a memory of 14 days of activity tracking and rechargeable via micro-USB cable with a duration of 8 hours in GPS mode and 7 days in basic smartwatch mode.
- ***Body+***: smart-scale (Withings, France; dimensions: 325 x325x23mm; weight: 2100g) rechargeable via AAA batteries with a battery life up to 18 months.
- ***iHealth Air***: smart-pulse oximeter (iHealth Labs Inc, USA; dimensions: 62x33x28 mm; weight: 42g) rechargeable micro-USB cable with a duration on a time scale of weeks.

As shown in Figure 2 the smart home devices are:

- ***Motion Sensors***: three sensors (Philips, The Netherlands; height: 5.5cm; length: 3 m; width: 5.5cm; net weight: 65g) for motion detection, temperature and level of lighting rechargeable via AAA batteries.
- ***Smart Bulbs***: sensor (Philips, The Netherlands; height: 11.8cm; width: 6cm; weight: 65g) for lighting adjustments including color, brightness and color temperature with an average life up to 25 years.
- ***Aqara Temperature Device***: three sensors (Xiaomi, China; dimensions: 36x36x9 mm, weight: 110g) for temperature, humidity and pressure with a battery life up to 2 years.

A total of 7 sensors, with a Raspberry Pi computer with a micro-SD card and a Zigbee adapter, constitute the home automation component of SB platform. It results fully compatible with any IoT platform, and it can be successfully linked with major AI technologies available on the market including Amazon Alexa, Google Assistant and Apple Homekit.

The parameters collected by personal devices of SMART BEAR are summarized in Table II while the measurements provided by home sensor devices of SB platform are shown in the Table III.

TABLE II. PERSONAL DEVICES PARAMETERS

Device	Parameters
Phonak Marvel-50	Duration of active use [hours or minutes]
	Average duration of active use per day [hours or minutes]
	Duration of exposure at environmental noise levels per day and per week [minutes]
	Percentage of active use in soft/medium/high intensity sounds [percentage]
Body+	Body weight [kilogram or pound]
	Body muscle mass [kilogram or pound]
	Body bone mass [kilogram or pound]
	Body fat mass [kilogram or pound]
	Body fat free mass[kilogram or pound]
	Body fat ratio [percentage]
BPM Core	Diastolic Blood Pressure [mmHg]
	Systolic Blood Pressure [mmHg]
	Heart Rate [beats per minute]
	ECG signal [μV, time series]
Thermo	Body Temperature [Celsius or Fahrenheit]
	Skin Temperature [Celsius or Fahrenheit]
iHealth Air	Blood oxygen saturation [percentage]
	Pulse rate [beats per minute]
Garmin VivoSport	Number of steps [dimensionless number]
	Distance traveled [meters]
	Calories burned through activity [kCal]
	Calories burned by Basal Metabolic Rate [kCal]
	Intensity Minutes [minutes]
	Duration of vigorous/moderate/low activity [seconds]
	Floors climbed [dimensionless number]
	Average heart rate on last 7 days [beats per minute]
	Average heart rate at rest [beats per minute]
	Sleep quality [label]
	Sleep duration [seconds]
	Time spent in deep/light/REM sleep [seconds]

TABLE III. SMART HOME DEVICES MEASUREMENTS

Devices	Measurements
Motion Sensors + Smart Bulbs + Aqara Temperature Device	Room Light Intensity [illuminance]
	UV Index [integer]
	Outdoor/Room Temperature [Celsius or Fahrenheit]
	Outdoor/Room Relative Humidity [percentage]
	Outdoor/ Room Atmospheric pressure [hectoPascal]
	Wheather conditions (i.e. wind speed, wind direction, rain volume, snow volume and visibility)

The main components of SMART BEAR cloud are the database and its underlying information model, the clinical repository interfaces, the big data engine, the synthetic data generation element, and the analytics. Therefore, a strong semantic underpinning and a standards-based data representation with comprehensive and homogeneous datasets are leveraged to ensure efficient use of data for implementing high-quality and personalized interventions improving QoL. In particular, the HL7 FHIR standard is used to provide the specification of the data model, and FHIR resource profiles are employed to define constraints and extensions to the FHIR model capturing the required information in a standard structure and with rich semantics. A federated and active learning approach is applied for the analytics since it allows training algorithms across multiple systems or devices holding local datasets without the need of exchanging data. That ML approach allows preserving data privacy while enjoying large-scale aggregation benefit. Moreover, it enables an optimization of local models while maintaining a high performance for the global model as asynchronous approach. Lastly, to ensure an accurate health monitoring and the consequent appropriateness of interventions provided by the platform, the quality of collected data is assessed in each training rounds and validation loops are introduced to detect any performance degradation and mitigate any performance drift.

V. SMART BEAR SCENARIOS & INTERVENTIONS

Based on the data gathered by personal and smart home devices, the cloud-level data analysis and the guidance of clinical staff, SB platform supports interventions that may help the management of every targeted medical condition. In particular:

HL. Interventions/notifications are aimed to *a*) increase the usage time of hearing aids (HA), *b*) obtain higher satisfaction with HA usage, and *c*) reduce the number of visits to the audiologist for HA fine-tuning.

CVDs. Interventions/notifications are aimed to *a*) increase the adherence to the therapy, *b*) improve older adult's outcomes and *c*) help in managing medical condition.

CIs. Interventions/notifications are aimed to *a*) meet the compliance with the clinician's recommendations (e.g., training time), *b*) assess factors that are likely to influence the progression of cognitive impairment, and *c*) improve older adult's outcomes as a result of interventions.

MHIs. Interventions/notifications are aimed to *a*) enhance user's motivation in physical activity, cognitive games and socializing, *b*) ensure a better compliance to daily activities, and *c*) improve older adult's outcomes and sleep parameters as a result of interventions.

BDs. Interventions/notifications are aimed to *a*) increase the adherence to the therapy and training, *b*) assess factors that are likely to influence the disease outcomes and the falling risk, and *c*) improve older adult's outcomes and as a result of interventions.

For every above medical condition, comorbidity and combination of them, existing medical guidelines and data stored in both FHIR and non-FHIR databases are considered, and clinician-driven interventions are generated by SB platform for each user individually. Such interventions, that may be adjusted based on the assessment of individual's actions in response to the treatment, are then transmitted via a security component to the older subject. The latter can access the own list of interventions through SMART BEAR mApp where, each intervention is displayed with its related medical conditions, priority level (i.e., Low, Medium, High), status (i.e., Active, Inactive), and notification text (e.g., "Repeat the measurement"). A version of the mApp is also available for the clinician that can thus insert new useful measurements, monitor the user's health status, and verify whether an intervention was successfully completed.

VI. CONCLUSIONS

In the era increasingly devoted to the promotion of "ageing well" and personalized medicine, SB platform represents a valuable tool to improve seniors' QoL. Firstly, this project will explore if the SB platform is an ideal tool for the continuous health monitoring of older adults with one or more than one medical conditions (e.g., HL, CVDs, CIs, MHIs, and BDs). Secondly, the SB project will plan and propose personalized interventions promoting an independent and healthy lifestyle. This is a project of Horizon 2020 that started in 2019 and is still ongoing. Therefore, all the results will be presented after the project's completion in 2023.

ACKNOWLEDGMENT

The project has received funding from the European Union's Horizon 2020 research and innovation program under grant agreement No 857172 – SMART BEAR project. No conflict of interest is declared.

REFERENCES

[1] R. Suzman and J. Beard, "Global health and aging", NIH Publ., vol. 1, no. 4, pp. 273-277. 2011.

[2] WHO, Active Aging: A Policy Framework, 2002.

[3] R.J. Scheidt, D.R. Humpherys, and J.B. Yorgason, "Successful aging: what's not to like?", J. Appl Gerontol, pp. 277-82, 1999.

[4] F. Sadoughi, A. Behmanesh, and N. Sayfouri, "Internet of things in medicine: A systematic mapping study", J. Biomed. Inform, pp. 103, 2020. doi: 10.1016/j.jbi.2020.103383

[5] S. V. P. Darcini, D. P. Isravel, and S. Silas, "A Comprehensive Review on the Emerging IoT-Cloud based Technologies for Smart Healthcare", pp. 606-611, 6th ICACCS, 2020.

[6] V. Bellandi, et al., "A design methodology for matching smart health requirements", CCPE, pp. 1–16, 2020.

[7] GBD Collaborators, "Global, regional, and national incidence, prevalence, and years lived with disability for 310 diseases and injuries, 1990–2015: a systematic analysis for the Global Burden of Disease Study 2015". Lancet. pp. (388): 10053, 2016.

[8] B.R. Rutherford, K. Brewster, J.S. Golub, A.H. Kim, and S.P. Roose, "Sensation and Psychiatry: Linking Age-Related Hearing Loss to Late-Life Depression and Cognitive

Decline", American J. of Psychiatry, vol. 175, no. 3, pp. 215-224, 2018.

[9] WHO, CVDs Key Facts, 2017.

[10] M. Moryś, M. Pąchalska, J. Bellwon, and M. Gruchała, "Cognitive impairment, symptoms of depression, and health-related quality of life in patients with severe stable heart failure", Int J Clin Health Psychol, vol. 16, no. 3, pp. 230-238, 2016.

[11] F. Jessen et al., "A conceptual framework for research on subjective cognitive decline in preclinical Alzheimer's disease", Alzheimers Dement, vol. 1, no. 6, pp. 844-52, 2014.

[12] D.L. Sturnieks, R. St George, and S.R. Lord, "Balance disorders in the elderly", Neurophysiol Clin, vol. 38, no. 6, pp.467-78,20

A Survey on AR/VR Games for Mental and Physical Public Health

1st Yu Fu
Department of Computer Science
Blekinge Institute of Technology
Karlskrona, Sweden
yu.fu@bth.se

2nd Yan Hu
Department of Computer Science
Blekinge Institute of Technology
Karlskrona, Sweden
yan.hu@bth.se

3rd Veronica Sundstedt
Department of Computer Science
Blekinge Institute of Technology
Karlskrona, Sweden
veronica.sundstedt@bth.se

***Abstract*—The use of augmented and virtual reality (AR/VR) has recently been increased in various people's work and life. The combination of technology with digital games have recently received much attention. The 2020 Augmented and Virtual Reality Survey Results report by Perkins Coie, XR Association, and Boost VC, pointed out that immersive techniques are leaving from laboratory research and theoretical applications and growing up as fully mature and income-generating marketplaces. Except for games and entertainment, the most common area reported was healthcare and medical devices from immersive technologies, followed by education. The survey presented in this paper explores the usage, perceived benefits and problems of AR/VR games for mental or physical health. It also highlights future features that users would see as relevant in these applications. Through an online questionnaire, we found examples of used AR/VR games and their usage frequency, the user's perception of their mental or physical health benefits and problems, and the prospects to physical and mental health for AR/VR games. This study is helpful for researchers and developers both in computer science and medical science. It provides people with examples of existing consumer-level products, their advantages and disadvantages, and AR/VR games' requirements relevant to mental and physical health. In future work, the survey sample size would be extended to collect additional user experience data and further explore the challenges and opportunities of such techniques for public health.**

***Index Terms*—Virtual reality, augmented reality, public health, games, survey**

I. Introduction

With the increased development of AR/VR of technologies, as shown in Fig 1, more research focuses on applications of the techniques and their evaluation. There is an increasing amount of consumer-level applications entering the lives of the public. Among them, AR/VR games have received much attention. Such games can provide fun and interest and bring a rich experience including serious objectives [1], such as benefits for users' mental and physical health.

A. Aim and Research Questions

From the perspective of the applied and evaluated techniques, this study explores the usage situation of AR/VR, thereby finding further challenges and opportunities and ideas for future studies. Thus, the main research objectives of this article are to explore the usage of the AR/VR games while bringing entertainment effects, what impact do these games have on the public's physical and mental health; and how the public's future expectations of AR/VR games could affect their physical and mental health. To address the aim, this study conducted an online questionnaire for the public to find potential perceived benefits and problems of VR/AR games for mental and physical health. The usage data regarding how often the AR/VR technology was used in addition to what game applications were tested were gathered from volunteers. The participants were also asked to imagine future features that these applications could incorporate. As such, the main research questions were as follows:

- RQ1: What is the usage experience of VR/AR games of the public?
- RQ2: What are the publics' perceived benefits and problems of VR/AR games for mental and physical health?
- RQ3: What features would the public like to see in future VR/AR games for mental and physical health?

B. Background

The following paragraphs present the definitions and relevant concepts of a game and AR/VR. Game is a rule-based system with variable and quantifiable outcome based on the assigned different values [1]. The game result relies on the players' effort [1] and the degree of understanding of the rules. A game can improve motivation and engagement for special purposes [2], not only for fun and entertainment but also for serious objectives [1].

As shown in Fig 2, mixed reality (MR) covers the whole range from the real world to the virtual environment, AR and augmented virtuality (AV) is the continuum between the natural world and the virtual world, where AR is closer to reality. Like a bridge from real to virtual, AR merges the physical and digital worlds, which is a three-dimensional technique, adds or removes computer-generated digital objects in the real world [3], thereby improving the sensory perception of the real environment by the content layer of information [4].

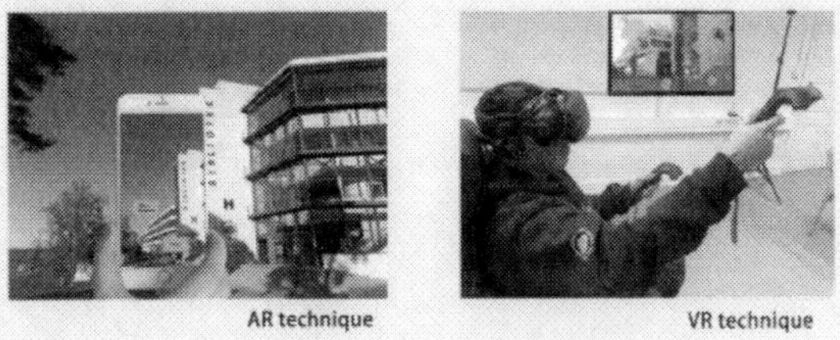

Fig. 1. An example of AR/VR techniques.

AR is not limited to a particular display technique nor the sight sense [4]. AR could operate on three main displays: head-mounted displays (HMD), handheld displays, and spatial displays, such as Microsoft Hololens, Google Glass; smartphone, tablet PC; and spatial augmented reality (SAR) by using video-projectors [5]. AR could also apply to all senses (including smell, touch, and hearing), such as sensory substitution application, which substitutes or augment missing senses for users. For example, AR could help blind and visually impaired users through audio cues and help deaf users through visual cues [6]. The AR technique is based on having support for three aspects: real objects tracking tools, information processing hardware and software, and display devices [3], [6].

In Milgram's continuum of real-to-virtual environments statement, as shown in Fig 2 [7], VR is the other end of the coordinate axis relative to the real world. VR was first introduced in the 1960s, which was described as a window, in which users could perceive and interact with a virtual world from the window as if looking, feeling, and sounding real world [8]. The definition of VR is varying, such as an immersion display technique with 3D models, which allow users to real-time interact with the models [9]; an advanced digital human-machine interface in virtual worlds [10]; a 3D immersive environment created by a computer [11], or to simulate physical presence and sense in a virtual environment to reproduce users' sensory experiences [12]. Overall, three main aspects are key here: immersion, interaction, and perception.

Based on the level of immersion, VR could be classified as non-immersive, semi-immersive, and immersive [13]. Different displays can be used to achieve different levels of immersion. The most immersive, such as VR glasses, HMDs, and CAVE, provide a completely immersive simulation experience for the VR applications. Equipment such as desktop, laptop, TV, or large screen supply the cheapest and simplest VR to recreate the real world. A semi-immersive VR is between the two above, which produce a 3D scene viewed on display using a perspective projection coupled to the head position of users, such as Fish Tank VR [13].

Moreover, the perception of VR also rely on audiovisual stimuli as a type of interaction, but some studies reported applications of multisensory VR systems, such as haptics (simple vibration, thermal, tactile, or kinesthetic), smell, and taste [14]. Furthermore, except for those mentioned above, which are relevant to output, the VR interaction also includes input devices. These range from keyboard, mouse, trackball, and joystick for lower immersive VR, to tracking devices for higher immersive VR, such as bend-sensing gloves (tracking hand movements/postures/gestures), pinch gloves (tracking the fingers movements), and users' physical movements trackers [13].

C. Related Work

Perkins Coie LLP, XR Association, and Boost VC conducted the fourth annual Augmented and Virtual Reality Survey Results report in early 2020 by 191 professionals [15]. Their report reviewed the development of the AR/VR industry in the past year and looked forward to its future. They pointed out that their interviewees represent strong optimism for the continued growth of immersive games and broader applications in other areas, such as healthcare, education, telecommunication, manufacturing, retail, and disaster preparedness. Among them, specifically, healthcare obtained more attention. They expected the healthcare industry in AR/VR would reach about 11 billion dollars by 2025. They also said more and more immersive technology applications were being recognised, such as simulated surgical training, remote diagnosis, pain management, palliative hospice care, and molecular level 3D visualisation of diseases. In other various industries, respondents also expected immersive techniques to improve operations and outcomes and enhance efficiencies. For example, such training for global staff could allow less cost for travel and lodging, as well as medical education and training with more opportunities for practice repetitions. The report claimed that immersive techniques appear to become a more significant part of people's daily lives worldwide. In addition, over 75% of the respondents expect that the total revenue of the AR market will eventually surpass that of VR due to the ubiquity of mobile devices that make it easier to use AR applications. On the other hand, the most need for improvement, which will impact users adoption of such techniques, is the continued upgrades of the device. Among them, 42% of the interviewees believed devices should be smaller and sleek, followed by 39% of participants' more comfortable [15].

There has also been a previous online survey in 2020 focusing on the possibilities and challenges with AR/VR/MR and gamification usage in healthcare [16]. Unlike the target participants and research aim of this study, in the previous work, we paid more attention to the staff, researchers, and students in the healthcare-relevant organizations, to discover the state-of-art of AR/VR/MR game applications/software in their working/studying. Based on 30 participants' answering, we explored the actual used digital games and AR/VR/MR applications applied in healthcare. This previous work collected the applications' general information (including name or feature, application purpose, target user, and use occasion), usage situation (covering use time and use frequency), user experience (advantages and disadvantages), obstacles to mass

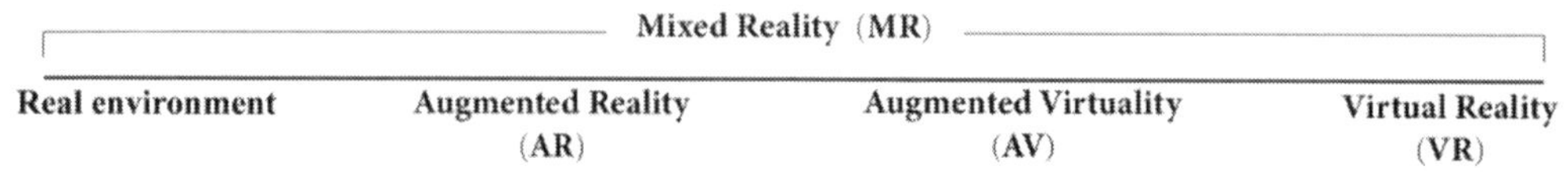

Fig. 2. The relationship of a real environment to virtual reality (VR) [7].

adoption for such techniques, and its future requirements and concerns.

In their answers, 15 software/applications were found. Five were AR/VR/MR games, the other five only involved AR/VR/MR techniques without any game element, and the remaining five were games without AR/VR/MR techniques. From the application purpose view, AR/VR/MR techniques were mainly applied to medical education and training (such as simulation CPR, model operation, 3D body map, virtual surgery, and X-ray imaging training) and mental health (such as stress management, anxiety relief, and for autism treatment). For the usage frequency view, only two participants mentioned it. One of them used a VR game (Deep) five hours per week. Moreover, interviewees claimed the advantages of the game in healthcare were in line with cognition and acceptance of children and stronger motivation than non-game activities. The reported problems were that of the high cost of software and hardware, the potential for being addictive, and not appropriately applied. Furthermore, the benefits of AR/VR/MR techniques in healthcare were the ability to change the setting for the patients/students (simulation CPR), intuitiveness, straightforward and easier to perceive (3D body map), more precise and valuable for the operation and better result (model operation), safe and allowed repetition (virtual surgery, X-ray imaging training), and better game experience (Deep). On the other hands, the disadvantages of AR/VR/MR were high cost and low quality in presentation. In addition, the interviewees mentioned five additional organisation requirements for AR/VR/MR games, such as rehabilitation, psychological and mental treatment, psychological diagnosis or psychotherapy, baby' taking care skills training, and games for elderly health. There were two more requirements mentioned for the AR/VR/MR techniques, being teaching and muscle relaxation.

Furthermore, there was a previous survey of AR/VR/MR technologies targeting another field, such as using the VR technique in higher education. From Cicek et al. work, we could see most of the interviewees (46/55, 84%) had experience with VR applications, 25 of them had such applications at home [17]. To compare a 2D display with HMD, they created several questions about the sense of the passage of time, immersion in the virtual world, and the opinions on the VR techniques. From their result, 77% thought VR systems could provide more information, 96% believed the visual stimuli of the VR applications was fascinating, and 60% claimed using the VR system would bring them to feel present in a virtual environment [17].

D. *Structure*

The rest of this paper organised as follow. Section II explains the research method, covering the research ethics, questionnaire design, procedure, participants, and analysis methods. Next, Section III presents the result based on 74 individual questionnaire answers, including the AR/VR games usage situation, benefits and problems of such games for mental and physical health, and the future features expectation of AR/VR games for mental and physical health. Section IV compares the outcomes with relevant previous work and explores the possible reasons behind the results. Finally, Section ?? gives a summary of the research and highlight its contribution and future work.

II. Method

This section presents the research method, including relevant aspects regarding research ethics, questionnaire design, structure and logic description, pilot study, procedure, participants and analysis methods.

A. *Research Ethics and Questionnaire Design*

Based on the General Data Protection Regulation [18] and Singer and Vinson's guidelines for practical ethics [19], we generated a research ethics explanation for the study, which was included as part of the informed consent form shown before the beginning of the online questionnaire. It also highlighted that no sensitive data was collected, there was no link to any person, as well as all data were confidential.

As shown in Fig 3, the questionnaire has three sections with 60 questions. Except for the Q1 and Q2, which are the consent form question and age group evaluation, the other 58 questions could be organised into two parts: (1) the usage of VR games and (2) the usage of AR games. To answer RQ1, each part included questions about VR/AR usage (Q3/Q32), use frequency (Q4/Q33), and application name (Q5/Q34). To answer RQ2, the questions (Q6/Q35) explored benefits for mental health and (Q7/Q36) benefits for physical health. To further address RQ2, the questions also mentioned VR/AR games advantages (Q8/Q37), and problems (Q9/Q38). Moreover, questions (Q31/Q60) were created for exploring RQ3 and imagined future features of VR/AR games for mental or physical health. Apart from Q31 and Q60, the other questions in each part were repeated three times (Q10-Q16, Q17-23, Q24-Q30, Q39-Q45, Q46-Q52, Q53-Q59) to deal with the participants who reported more than one VR/AR game user experience. If they did not have any more experience, they could directly skip the repeated questions. The mandatory questions were marked with a star.

The questionnaire were designed with single choice questions, 7-Likert scales, and open questions. For the first two, we quantified the options for descriptive statistics and statistical analysis. For the latter, we extracted and analyzed the high-frequency key words in the answers. Before the formal document survey, the researcher conducted a pilot study to test the usability of the questionnaire and its provided link.

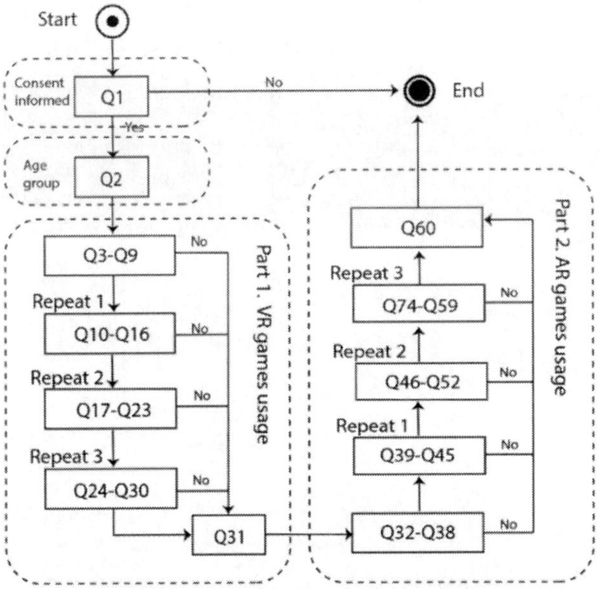

Fig. 3. The questionnaire structure and logic.

B. *Procedure and Participants*

The questionnaire was created in Microsoft Forms and spread by social media (LinkedIn, Facebook, Twitter, and WeChat). By mid-April (2021-04-15), there were 76 answers. Two declined to join this survey resulting in 74 effective obtained replies. The following analysis is based on the 74 answers. The age group of the participants was mainly concentrated within the range 25-30 years old (52), followed by the age group 18-24 (17). Four participants came from the age group 35-44, and only one from the 45-55 age group.

III. RESULTS

To answer the three RQs, this section organises and analyses the results in four subsections: the AR/VR games usage situation (III-A), perceived benefits (III-B) and problems (III-C) for mental and physical health, and the future imagined features of AR/VR games for mental and physical health (III-D).

TABLE I
THE AR/VR GAME USAGE SITUATION OUT OF POSSIBLE FOUR OPTIONS.

	Q3/Q32	Q10/Q39	Q17/Q46	Q24/Q53
VR	38	6	4	2
AR	21	2	1	1

TABLE II
THE USAGE FREQUENCY OF THE MENTIONED AR/VR GAMES.

(a week)	VR				AR			
	Q4	*Q11*	*Q18*	*Q25*	*Q33*	*Q40*	*Q47*	*Q54*
Less than once	35	4	4	1	16	1	0	0
2-3 times	3	1	0	1	2	0	0	0
Over 3 times	0	1	0	0	2	1	1	1

A. *Usage Situation*

Table I shows the number of "Yes" answers from Q3/Q32 (Do you have experience with VR/AR game applications?) and their three repeated questions (Q10/Q39, Q17/Q46, Q24/Q53). It can be seen that from the possible four times data could be collected, the VR technique (50 person-times in total answered yes) had been more used than the AR technique (25 person-times in total answered yes) out of the 74 volunteers of the questionnaire.

1) Application Purpose: From the positive answers, there were 23 applications mentioned. Among them, 17 were VR games, and the others were games with AR techniques. Based on the game genres, the 24 games could be grouped into eight types. As shown in Table III, the VR shooter game and AR adventure game were the most used game genres. VR shooter games (6) was also the genre with the most number of games mentioned in this survey, followed by VR adventure games (4) and simulation games (4). Only one game was mentioned in the AR adventure game category. However, this game was mentioned by ten of the participants. This outcome is quite natural since the game *Pokémon Go* previously has been reported as one of the most popular games with health benefits [20], [21]. *Pokémon Go* has also previously been evaluated in terms of physical and psychological benefits for dog owners [22].

2) Use Frequency: Although more than half of interviewees have VR/AR game experience, the use frequency of most games was lower than once per week. As shown in Table II, only six people mentioned they play the VR/AR games more than three times per week and the other six people 2-3 times per week.

B. *Perceived Benefits of AR/VR Games to Health*

Even little play still provides the user experience on health benefits. Fig 4 shows the statistics of the mentioned games' benefits on mental and physical health. Horizontal comparison, most of them believed the AR games benefit their mental (18/25, 72%) or physical (19/25, 76%) health. However, the percentages for VR games were lower. Twenty-nine person-times thought that the VR games had benefits for their mental health (29/50, 58%), and twenty-four person-times reported supported benefits for physical health (24/50, 48%). Vertical comparison, more than twice as person-times who disagree, there was 90 person-times thought that AR/VR games were good for physical and mental health. Among them, there were four more person-times (47 vs 43) who believed AR/VR games were beneficial to their mental health than their physical health.

TABLE III
OVERALL AMOUNT STATISTICS OF THE MENTIONED VR/AR GENRES AND GAMES.

	VR			AR		
	Genres	*Games*	*Name/Feature*	*Genres*	*Games*	*Name/Feature*
Action game	1	1	*Sword battle*	1	1	*Alipay's Ant Manor* AR hand exercises [1]
Shooter game	11	6	*Beat saber* [3], Shooting zombie [2], Shooting game [1], First-person shooter game [3], *Fruit Ninja* [1], *Gunfight* [1]	1	1	Dead lands [1]
Rhythm game	1	1	Dance game	0	0	null
Adventure game	4	4	*Moss* [1], Discovery game [1], Sword and magic [1], *Mage guard: the last grimoire* [1]	10	1	*Pokémon Go* [10]
Role-playing game	1	1	*Perfect World* [1]	0	0	null
Simulation game	5	4	Car racing game [2], Fire fighting simulation [1], *Roller Coaster* [1], Flight training [1]	0	0	null
Strategy game	0	0	null	2	1	*Onmyoji* [1]
Casual game	0	0	null	2	2	Take picture with dinosaur [1], Collecting "Fu" card [1]

Note: In the "Name/Feature" columns, the italic names are real games, whereas the other ones are only a type of game feature. The number in "[ ]" means the total number of participants mentioning such games.

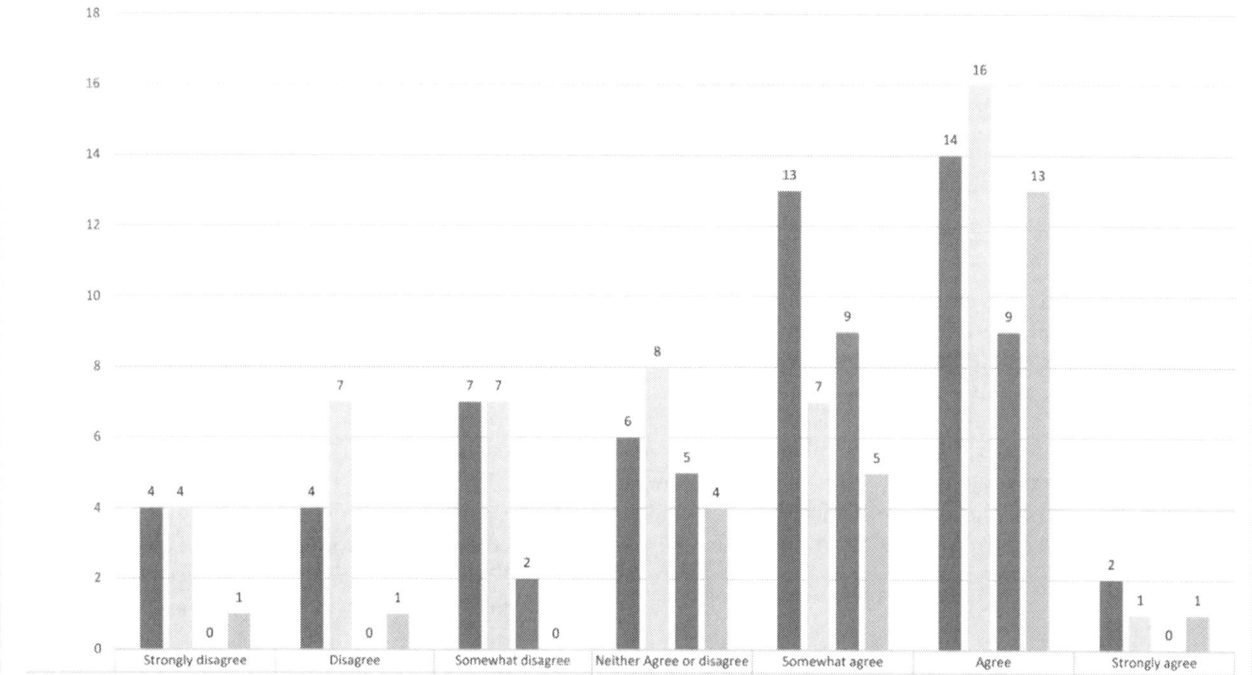

	Strongly disagree	Disagree	Somewhat disagree	Neither Agree or disagree	Somewhat agree	Agree	Strongly agree
VR Mental	4	4	7	6	13	14	2
VR Physical	4	7	7	8	7	16	1
AR Mental	0	0	2	5	9	9	0
AR Physical	1	1	0	4	5	13	1

Fig. 4. Distribution of the 7-point Likert scale results of whether AR/VR games benefits the mental or physical health (Strongly disagree-Strongly agree).

Before the formal data analysis of whether the mentioned games benefit mental and physical health, we explored the reliability and validity of the scales, by a Cronbach's Alpha (CA) test and Exploratory Factor Analysis (EFA) through SPSS [23]. The answers were transformed to numbers in the statistical analysis (strongly disagree 1, disagree 2, somewhat disagree 3, neither agree or disagree 4, somewhat agree 5, agree 6, strongly agree 7).

Due to that the CA results were over 0.6 (AR: 0.710, VR: 0.606), we could say our scales have internal reliability. Moreover, based on the result of the Kaiser-Meyer-Olkin measure (KMO) and Bartlett's test, in which the KMO of Sampling Adequacy was greater than or equal to .50 (AR: 0.50, VR: 0.50), and the significance value less than 0.05 (AR: 0.002, VR: 0.004), thereby we could say our data was suitable for the EFA [24]–[26]. From the results of the EFA; the Total Variance Explained results shown the cumulative variance contribution rate of the four questions (VR-M/AR-M: questions of whether VR/AR games benefit mental health, VR-P/AR-P: questions of whether VR/AR games benefit physical health) divided into two dimensions (AR-M/AR-P and VR-M/VR-P) was 78.18. Due to the value being over 60, which means the classification is suitable [26]. In addition, the results of the Rotated Component Matrix shows that each question only has one contribution value to these two dimensions, which proves that they contributed to the scale separately and cannot be deleted.

We first conducted the normal distribution test to select the statistical method for the difference relationship analysis. The data of the VR-M, VR-P, AR-M, AR-P answers and the total combined attitude of VR-C (for mental and physical together) was in line with the normal distribution, but the total attitude of AR-C (for mental and physical together) was not. Thus, to compare data groups with equal sample size and normal distribution (such as VR-M/VR-P, and AR-M/AR-P), we used the paired T-test (parametric test). However, for the analysis of the data group that had different sample size or not in the normal distribution (such as VR-M/AR-M, VR-P/AR-P, and VR-C/AR-C), we used the Wilcoxon test (non-parametric test).

For the two rounds of paired T-tests, the Sig. values were over 0.05 (VR-M/VR-P: 1, AR-M/AR-P: 1). Thus, the difference in the answer of VR-M and VR-P, as well as AR-M and AR-P, were not statistically significant. Thereby, we could say, facing the technology, AR or VR, the interviewees have similar attitude towards them whether benefits their physical and mental health. Moreover, for the three rounds of Wilcoxon test, the Sig. values were less than 0.05 (VR-M/AR-M: 0.017, VR-P/AR-P: 0.027, VR-C/AR-C: 0.018). Thus, the difference in the answers of VR-M/AR-M, VR-P/AR-P, and VR-C/AR-C was statistically significant. Based on the result, we could say, facing different health needs, physical or mental or both, the participants' attitudes towards different technologies were perceived different.

From the result of advantages in VR games, the participants mentioned some high-frequency keywords, such as immersive, fun (happy), relax (release), and real (realistic, realism). From the above keywords, the interviewees began to describe the user experience of VR games as a more stimulating environment, more interesting, reducing pressure, and pushing the attention only to the game. They also said VR games gave them a new and improved game experience to enjoy and discover a new place they never been to before. VR games could reduce the limitations of entertainment and the real danger of experience in a danger-like situation. Moreover, some participants believed that VR games could provide an online social community, making them feel less lonely. Furthermore, "exercise" and related words were said several times. Someone specifically claimed that VR games could motivate more exercise. Another pointed out that the VR game help to strengthen his/her body. Furthermore, the VR game can benefit reaction practice. Last but not least, a participant mentioned that the VR game could provide great visual effects.

For AR games, the most perceived advantages were around motivating the players to go outdoors, which is beneficial to their bodies and to see the outdoor environment, thereby enhancing mental and physical health. Moreover, some interviewees said, AR games do not need additional equipment and that it is easy to use. Furthermore, AR games were reported more novel and immersive compared to traditional games. Other participants pointed out that they could bring a good visual experience, interest, and happiness to the users.

C. Perceived Problems of AR/VR Games to Health

On the other hand, the disadvantages of VR games were mainly focused on the physical discomfort and risks with equipment. Many participants pointed out the dizziness and motion sickness problem caused by 3D vertigo. Another remark concerned negative impacts on the eyes. Some interviewees believed VR games were not good for the eyes. Except for the tired and uncomfortable eyes, they claimed that playing VR games for a long time could easier cause eyestrain and deepen myopia. Someone also mentioned being dazzled when he/she used the VR game. Moreover, the physical discomfort and risk also included neck pain and risk of falling. Furthermore, some interviewees said that when playing VR games, the player or the surrounding environment were at risk of injury or damage.

Another main problem came from the device. A participant mentioned the high cost of software and equipment. Some others pointed to that devices are too heavy, and this being the main problem affecting the user experience. Furthermore, due to the device, playing VR games was not easy at home. It needs enough space to layout equipment and play, which limited the playing environment. The cable impacted movement and easiness to twist and wind when playing the VR games as well. Moreover, some answers showed a negative impression on imaging. Someone claimed the VR device was not mature enough, and the imaging had bad details.

In addition to the above main issues, the respondents' negative impression of VR games lied in their impact on the real world. Some interviewees said that after a long time of playing, the VR game impacted the feeling of the true world, more than this, the influence also on the distinction between the real world and the virtual world. Besides, VR games were also reported to have similar problems as other games, such as addiction, wasting time, and reducing communication with friends face to face.

The AR games have reported usage problems as well. The same as for VR and other games, addiction was mentioned by some interviewees. Some participants compared the user experience of AR games with VR games. They said AR games were not as fun as VR games and could not make them immersed due to that the design is not so real. Moreover, they also claimed device problems, such as networking problems, the high challenge of their cellphone battery, and the limitation of the screens. Except for the above comments, the risk of traffic accidents was pointed out several times. Some respondents said that when playing AR games, they may ignore their surrounding environment. Furthermore, there was one interviewee who reported the AR technique to be seriously abused.

D. Proposed Future Features of AR/VR Games for Health

Some participants mentioned the sci-fi movie "Ready Player One". Adapted from the novel of the same name, it depicted much imagination about VR and reality. The interviewees imagined living realistically in the VR world like the movie's description. It imagined a high mix between the VR world and the real world, blurring the boundaries between realism in a virtual world and the virtual in the real world. The respondents' positive attitudes towards VR technology and demand for the realism of virtual objects are shown here. It is also reflected in keywords that appear in the answers many times, such as "pleasure", "enjoyment", "immersive", "simulation", and "real (realistic)". The interviewees also described their requirements and expectations of VR techniques and VR games into two main aspects: user experience and content offerings.

The most mentioned comments were about multi-senses, multi-user interaction, unlimited location of use, wireless, naked eyes VR from the user experience point of view. Several participants claimed they want future VR games to perceive senses, including gustatory, tactile, and olfactory, as the new interaction brings a more realistic and immersive experience. In addition, interaction with other players was also pointed out as important. Interviewees wanted to see and collaborate with other users. Furthermore, a VR should be possible to play without specific physical demands on the location. Some answers showed that the participants need to use it in a wide barrier-free venue, playing at home and without going outside. Except for the above, the expectations of VR games were to be more portable, easy to apply and to be cheaper. Users complained that helmet glasses are too heavy and that there is limited activity when having cables to consider. Wireless VR and naked eyes VR were mentioned several times. Moreover, someone also pointed out to have body control and not just the controllers using the hands.

From the content offerings point of view, many interviewees mentioned the VR games' application purposes in healthcare. As an aim for mental and psychological health, they pointed out that VR games could be applied for mental training, mental treatment, and psychological counselling. Especially for diseases and problems such as social barriers, autism, and Alzheimer. In addition, they claimed VR games could help to release pressure and promote or improve sleep. Facing physical health, participants mentioned that VR games could be used as a training and treatment method. Some of them believed it might help disabled people and expected it to recover muscle activation. Furthermore, several participants hoped VR games could support extreme games, sport and exercise, such as skydiving, VR gyms, live football games, and other interactive exercises. Except for directly health-relevant purposes, respondents mentioned aims of VR games' applications to be for education and learning (such as emergency handling for children), training (in schools, construction sites, and office buildings), team building, travel and tours, scene and memory reappearance, virtual meetings, record and observe, and for religious activities.

In addition to the purpose of the application, the answers also described the expectations for the functions and features. Some interviewees imagined VR games with biofeedback, such as perceiving individual moods or emotions and recommended games suitable for the current mood. These were reported to use, for example, advanced chips that could connect to the human brain, monitoring of heart rate detection, pulse detection, and other physical indicators. Moreover, when playing VR games, participants wished to break the sitting or standing playing mode and being able to run and exercise, and provide the leg movement system, in-situ movement. Furthermore, more freedom and interaction and less distance between non-player characters (NPCs) in video games and players were also pointed out for an improved experience. However, negative opinions could also be found in the answers, which involved addiction and harm to the eyes as other types of digital games. In addition, someone said VR is not a very mature technique.

As the expectations of VR games, the participants described their vision of the future for AR games in healthcare and their potential application purposes and features. Except for the requirements of training and treatment for mental and physical health, such as depression, stress relief, sitting a long time, eye protection, and fitness/sport/exercise, they also mentioned that AR games could help education and communication in health and telemedicine (remote surgery). For example, someone said AR games might be used to introduce the human body, how it works, what conditions can cause injuries, and learn things about healthcare. Furthermore, it might also help to study human anatomy. The 3D structure of the human body could be generated by an AR game based on computed tomography (CT) scan or other tools. Moreover, doctors and patients could use AR games to better communicate, such as for therapies, pills, and symptoms. In addition, there were also some mentioned application purposes not directly relevant to healthcare, such as online shopping, virtual meetings, navigation and map exploration, movies, record and observe, product design, travelling, learning and improving knowledge and skills, and experiencing different life types.

About AR functions/features and devices, the results are also similar to VR games. Interviewees said light AR glasses might attract players more. They also pointed out that if an AR game was more open and free, with better realism and immersion, the interaction and connection with the real world would be improved. Moreover, a better sensory experience with tactile feedback was also required. On the other hand, participants also held negative opinions on the use of AR games in health. One of them reported that AR could not be used in any healthcare project. Some others mentioned problems of violence and that violent games being too real could potentially increase the player's violent tendency and make it harder to distinguish between the virtual and real world. One interviewee said it is necessary to protect teenagers from such problems. Another person complained that AR games were not vivid enough.

IV. Discussion

It is noticed that all participants were in the age range 18-55, and most of them were under 30 years old. Although the adoption of smartphones, the Internet, and social media has risen amongst seniors over the last 15 years, it is still relatively low compared to young people. A 2017 survey in the United States [27] showed that 40% of senior citizens now own a smartphone. Even this number was more than double the amount in 2013, but still much lower than the smartphone ownership rate of young people. Moreover, the separate Internet and broadband use were different in various demographic groups. The report pointed out it was impacted by educational attainment and family income. Compared with households with an annual income of $75,000, the Internet usage rate of the elderly in households with an annual income of less than $30,000 has dropped by nearly 50% 94% vs 46%. Compared with those with a college degree or below, more college graduates use the Internet (92% vs 49%). Furthermore, the report also claimed 34% of elderly who are over 65 uses social media such as Facebook or Twitter. Even though the number was 6% higher than in 2013, it is significantly smaller than the general population. The majority of the elderly did not use social networking, and the younger elderly more possible use social media than the elderly. The lower use rates of the smartphone, the Internet, and social media of the elderly was both the limitation of the research method (online questionnaire spread by social media) and the challenges of AR/VR techniques. Due to an increase in elderly users of the technology, the positive view for the future could allow a growth in opportunities for this age group as well. Based on these results the authors also think it would be relevant in the future to find novel AR/VR game solutions, for mental and physical health, that look into possible collaborations between young and elderly, such as proposed in [28].

In the data analysis, we found one answer that confusing AR and VR technologies. Someone thought "Pokémon Go" was a VR game. It might be just a mistake, but it also could due to the vague understanding of such techniques. It could reflect that there is still room for AR/VR techniques application and popularisation to a certain extent. Facing AR/VR techniques, the participants have similar opinions for such applications benefits to their health, but with different attitudes for applying AR/VR on different health issues. Although far more VR games were mentioned than AR (17 vs 6), a single AR game had far more users than a single VR game (10 vs 3). This confirmed to a certain extent that the AR technique had greater development potential, mentioned in the related work, due to the popularity of mobile devices that can run AR applications. Some interviewees had a negative attitude towards AR/VR games and its benefits to physical and mental health, but far more people held the opposite opinion. As shown in the results III-B, several advantages could be seen. For example, VR could provide a more interesting and attractive game experience with fewer limitations and dangerous risks. It could also stimulate sports' motivation and being used to enjoy social communication online. In addition, AR games do not require additional equipment apart from a phone and could be encouraged to go outdoors. Those are some of the opportunities for AR/VR game development, especially for mental and physical health. However, the challenges should be paid attention to as well. Specifically, the device/equipment problems in cost and still be perceived as uncomfortable are shown in III-C.

Comparing the result of the related work and this study, we could see the similarities of the future requirements of AR/VR games, like mental, physical and psychological health, education, and learning. It is worth noting that the training, healing, and relaxation for psychological and mental health were the most mentioned in the related work and the survey conducted in this paper. The previous research and this study also have similar results of relatively low use frequency of AR/VR software applications and games. This was similar for healthcare-relevant staff or students in previous work and the public as reported in this survey. This shows that AR/VR techniques still have room for development and popularisation. In a comparison between the studies, there are also similar results in the benefits and problems of AR/VR/MR and games. For example, both mention more cost-effective devices and software, the potential of being addictive, space for hardware equipment to improve, stronger motivation to continue its use, safety, portability, and an improved experience.

V. Conclusion and Future Work

This paper has presented an online questionnaire, based on 74 interviewees, exploring and evaluating the usage situation, the impact of public physical and mental health, and future features in AR/VR games. Most interviewees had experience in AR/VR games, but their practical use frequency was low. Among the 23 AR/VR games mentioned, the number of VR games were higher than AR games. However, the number of users of a single game is far greater in AR than in VR. More than 70% of participants believed that AR games are good for the body and mind, but the data for VR games was around 50%. Interviewees mentioned many of the benefits, problems, features, and functions requirements in the future development of AR/VR games. These could provide references to researchers and developers about what is worthy of attention to improve the technology and applications. Future work will also be extending the sample size, especially to an elderly group (over 55 years old), to better know their opinions of AR/VR games for mental and physical health. As such, we can better understand their specific requirements for such techniques, thereby enriching and perfecting the results and making them more universal.

REFERENCES

[1] T. Lima, B. Barbosa, C. Niquini, C. Araújo, and R. Lana, "Playing against dengue design and development of a serious game to help tackling dengue," in *2017 IEEE 5th International Conference on Serious Games and Applications for Health (SeGAH)*, pp. 1–8, IEEE, 2017.

[2] S. Deterding, D. Dixon, R. Khaled, and L. Nacke, "From game design elements to gamefulness: defining gamification," in *Proceedings of the 15th international academic MindTrek conference: Envisioning future media environments*, pp. 9–15, ACM, 2011.

[3] R. Azuma, Y. Baillot, R. Behringer, S. Feiner, S. Julier, and B. MacIntyre, "Recent advances in augmented reality," *IEEE computer graphics and applications*, vol. 21, no. 6, pp. 34–47, 2001.

[4] R. T. Azuma, "A survey of augmented reality," *Presence: Teleoperators & Virtual Environments*, vol. 6, no. 4, pp. 355–385, 1997.

[5] R. R. Bimber, Oliver and M. Inami, *Spatial Augmented Reality SIGGRAPH 2007 Course 17 Notes*. SIGGRAPH, 2007.

[6] J. Carmigniani, B. Furht, M. Anisetti, P. Ceravolo, E. Damiani, and M. Ivkovic, "Augmented reality technologies, systems and applications," *Multimedia Tools Appl.*, vol. 51, p. 341–377, Jan. 2011.

[7] P. Milgram and F. Kishino, "A taxonomy of mixed reality visual displays," *IEICE TRANSACTIONS on Information and Systems*, vol. 77, no. 12, pp. 1321–1329, 1994.

[8] I. E. Sutherland, "A head-mounted three dimensional display," in *Proceedings of the December 9-11, 1968, fall joint computer conference, part I*, pp. 757–764, 1968.

[9] H. Fuchs and G. Bishop, "Research directions in virtual environments," 1992.

[10] J. N. Latta and D. J. Oberg, "A conceptual virtual reality model," *IEEE Computer Graphics and Applications*, vol. 14, no. 1, pp. 23–29, 1994.

[11] M. Kera, H. Pedrini, and F. Nunes, "Ambiente virtual interativo com colisão e deformação de objetos para treinamento médico," *Revista de Informática Teórica e Aplicada*, vol. 18, no. 2, pp. 205–233, 2011.

[12] X. Tong, D. Gromala, A. Amin, and A. Choo, "The design of an immersive mobile virtual reality serious game in cardboard head-mounted display for pain management," in *International Symposium on Pervasive Computing Paradigms for Mental Health*, pp. 284–293, Springer, 2015.

[13] P. Cipresso, I. A. C. Giglioli, M. A. Raya, and G. Riva, "The past, present, and future of virtual and augmented reality research: a network and cluster analysis of the literature," *Frontiers in psychology*, vol. 9, p. 2086, 2018.

[14] M. Melo, G. Gonçalves, P. Monteiro, H. Coelho, J. Vasconcelos-Raposo, and M. Bessa, "Do multisensory stimuli benefit the virtual reality experience? a systematic review," *IEEE Transactions on Visualization and Computer Graphics*, 2020.

[15] X. A. Perkins Coie and boost VC, "2020 augmented and virtual reality survey report," 2020.

[16] Y. Fu, Y. Hu, V. Sundstedt, and C. Fagerström, "A survey of possibilities and challenges with ar/vr/mr and gamification usage in healthcare," in *HEALTHINF 2021 - 14th International Conference on Health Informatics*, IEEE, 2021.

[17] I. Cicek, A. Bernik, and I. Tomicic, "Student thoughts on virtual reality in higher education—a survey questionnaire," *Information*, vol. 12, no. 4, p. 151, 2021.

[18] P. Voigt and A. Von dem Bussche, "The eu general data protection regulation (gdpr)," *A Practical Guide, 1st Ed., Cham: Springer International Publishing*, 2017.

[19] J. Singer and N. G. Vinson, "Ethical issues in empirical studies of software engineering," *IEEE Transactions on Software Engineering*, vol. 28, no. 12, pp. 1171–1180, 2002.

[20] M. N. Kamel Boulos, Z. Lu, P. Guerrero, C. Jennett, and A. Steed, "From urban planning and emergency training to pokÃ©mon go: applications of virtual reality gis (vrgis) and augmented reality gis (argis) in personal, public and environmental health," *International journal of health geographics*, vol. 16, no. 1, pp. 7–7, 2017.

[21] J. Hamari, A. Malik, J. Koski, and A. Johri, "Uses and gratifications of pokemon go: Why do people play mobile location-based augmented reality games?," *International Journal of Human Computer Interaction*, vol. 35, no. 9, pp. 804–819, 2019.

[22] L. Kogan, P. Hellyer, C. Duncan, and R. Schoenfeld-Tacher, "A pilot investigation of the physical and psychological benefits of playing pokmon go for dog owners," *Comput. Hum. Behav.*, vol. 76, p. 431–437, Nov. 2017.

[23] L. L. Chan, N. Idris, *et al.*, "Validity and reliability of the instrument using exploratory factor analysis and cronbach's alpha," *International Journal of Academic Research in Business and Social Sciences*, vol. 7, no. 10, pp. 400–410, 2017.

[24] H. F. Kaiser, "An index of factorial simplicity," *Psychometrika*, vol. 39, no. 1, pp. 31–36, 1974.

[25] B. Williams, A. Onsman, and T. Brown, "Exploratory factor analysis: A five-step guide for novices," *Australasian journal of paramedicine*, vol. 8, no. 3, 2010.

[26] A. G. Yong, S. Pearce, *et al.*, "A beginner's guide to factor analysis: Focusing on exploratory factor analysis," *Tutorials in quantitative methods for psychology*, vol. 9, no. 2, pp. 79–94, 2013.

[27] M. Anderson and A. Perrin, "Technology use among seniors," *Washington, DC: Pew Research Center for Internet & Technology*, 2017.

[28] E. T. Khoo, A. D. Cheok, T. H. D. Nguyen, and Z. Pan, "Age invaders: Social and physical inter-generational mixed reality family entertainment," *Virtual Real.*, vol. 12, p. 3–16, Mar. 2008.

Linked Care - Information Transfer in Mobile Care and Support

Kathrin Mühlhauser, Franz Werner
Department Health Sciences, FH Campus Wien
Vienna, Austria
e-mail: kathrin.muehlhauser@fh-campuswien.ac.at, franz.werner@fh-campuswien.ac.at

Elisabeth Haslinger-Baumann, Theresa Galanos, Doris Zeidler
Comepetence Center for Applied Nursing Research, FH Campus Wien
Vienna, Austria
e-mail: elisabeth.haslinger-baumann@fh-campuswien.ac.at, theresa.galanos@fh-campuswien.ac.at,
doris.zeidler@fh-campuswien.ac.at

***ABSTRACT*—Mobile care professionals are facing immense workload, less time, and little recognition. They often have to finish their documentation work in leisure time and do not get enough information from the other health professionals. Information and Communication Technology (ICT) support can bring significant improvements for home care and the involved care giving parties. To increase its potential and operating range, the project, named Linked Care (LICA, consisting of the first to letters of both words), develops IT systems that address and support practice-oriented challenges, increase digital data availability and reduce the work-related burdens of health care professionals as well as clients/patients and their involved support system. Experiences with existing IT systems plus the current existing Austrian Electronic Health Record System (ELGA, an acronym of Elektronische Gesundheitsakte) serve as a starting point to enable a continuous information supply in mobile care and support within the project. The solution will provide interdisciplinary support and the connection of relevant participants in the care process, offering a new type of responsive user interface. The LICA-project investigates the end-users' needs and specifies the processes in workshops with local representatives and regional political and administrative stakeholders. In this way, the LICA-solution allows for the exchange and evaluation of care data via standardized interfaces throughout Austria, in consideration of the socio-economic regional and national environments. The project will create an ELGA-compatible software portal that can be linked to existing systems. Large-scale tests will be carried out in different federal states with at least 60 clients/patients (persons who are in the need of homecare) while implementing/using different survey methods to collect data. The consortium parties will ensure the development of the solution as a business model. Requested findings include feedback on software usability, application and affordability for the care giving and care receiving target group. The generated Care Summary creates new possibilities for data exchange within the health care sector.**

Keywords-IT documentation systems; communication systems; health care and nursing; interdisciplinary research

I. INTRODUCTION

Nurses experience many burdens and challenges in their daily work. Documentation is often described as particularly stressful [3]. According to estimates, about 30% of working hours are used for documentation. This time cannot be spent with clients/patients, which in turn can lead to nurses finishing documentation work in their free time to have more available hours for direct client-based care [5]. The increasing number of people in need of care leads to an increase not only in the objectively measurable but also in the subjectively experienced care effort. As the complexity within the formal and informal care support network is increasing, the need for adequate client-related coordination between the involved parties is also growing. A continuous and detailed client documentation is the starting point of client related coordination and communication. However, the organization of client documentation in the care system is not only challenging in terms of effort, it also shows deficiencies and complications regarding the information flow and exchange between the different formal and informal care giving/providing parties. Adequate documentation is indispensable for high-quality nursing and medical care, as well as interdisciplinary cooperation. Due to the dual financing system of medical care, nursing care and homecare, the resulting responsibilities may be different in nature and "not necessarily conducive to the efficiency of the systems" [2]. Furthermore, the lack of interdisciplinary exchange is closely related to financial and time constraints: only 60% of all mobile care service providers cooperate with professional groups outside their own services [2]. On the one hand, the heterogeneity of the documentation and communication systems offered by different providers complicates the communication connection/link within the different mobile services/software solutions. On the other hand, the non-existence of an obligatory uniform solution forms a massive barrier to continuous information transfer. Sophisticated services, which are needed to support recent technological innovations, have not been provided while creating data gaps in all areas of social welfare and health care. This causes discontinuance and translation problems within software solutions.

So far, the record validity was and still is questionable due to existing language barriers between the carer's first language and the available application languages. Comprehensive documentation is particularly important. As an example, specific groups were identified, that are not familiar with professional care within their households (e.g., people with migration background, people with mental illnesses, etc.). The lack of affordability in the presence of social disadvantage or the lack of information of particularly vulnerable groups, for whom care is mainly provided informally or within the family. The situation described above points out the need for comprehensive documentation and communication that combines several information clusters. Therefore, it contributes to a subjectively facilitated situation and to an objectively increased efficiency [2]. It also advocates the improvement of communication and interface management based on detailed surveys.

The aim of this project is to enable a technology-based cooperation between clients/patients as well as their support system and professional caregivers, nursing professionals, doctors, therapists, and pharmacies while using an efficient, secure and low-threshold digital tool, offering optimal IT support.

This paper will give an insight into the LICA-project; starting with the 'State of the Art' in Section 2, followed by Section 3, 'Method', and finally describing the 'Results' as well as the 'Conclusion and Further Works' in Section 4 and 5.

II. State of The art

Care network services exist all over the world. However, they neither include all necessary participants (medical care providers, pharmacies, care organizations, etc.) nor do they function on a digital level. Once integrated into one organization other services needed such as visiting services, home services and nursing care have to be organized by the employees [7].

In the EU and the USA there have been some digital developments in this field. Siemens eHealth Solutions [10] has created an electronic health network that brings together clients/patients, care teams and medical doctors to exchange relevant data. This system was designed for the clinical setting and not for the homecare sector. Furthermore, it does not include clients involved relatives or therapists. In the USA, the My HealtheVet portal [9] is available for Veterans by Veterans Health Administration (VHA) to renew prescriptions, organize doctor's appointments, contact health care teams, and retrieve information.

Online Care is an US-platform for clients/patients and healthcare providers that is oriented towards GPS (or postcode) entries and enables corresponding connections. In Europe, cross-border healthcare data flow is already established, especially in the "eHealth Digital Service Infrastructure" [8]. Patient Summaries, ePrescriptions and eDispensation, are currently being rolled out. In Austria, the electronic health record system ELGA uses the same IT and technology standards, to create a feasible connection. In addition to numerous other initiatives, the EU is also striving for an "EU Health Data Space" to make available data more accessible for practical use within the healthcare system.

III. Method

The research project started in April 2021 und will last until 2025. Five end-user partners, five technology partners and three scientific/research partners are included, whereby one partner is accountable for the information privacy protection mechanisms.

All required technical functionalities of the product will be identified and described by the team of developers and the potential end-user, while implementing the user-centered design approach [4], which is already currently running. Moreover, a mixed method approach, supported by a profound literature research, is used to identify, and approach the target groups in question as well as implement the appropriate methods for the various target groups and settings. By now, methods like focus group interviews, one-on-one interviews as well as research diaries are applied. The use of a participatory approach [6] enables the involvement of people and groups that are difficult to reach (e.g., 24-hour care but also very vulnerable people). All research methods are applied low threshold and in a culturally sensitive manner.

Additional stakeholders from the regional and national environment are involved to find solutions for specific issues. Over the duration of the project, goal-oriented, cross-disciplinary, and cross-role networks of individuals and organizations will be created. Sustaining and continuing these networks after the end of the project, represents one of the long-term goals. Acceptance factors are defined by the experts for all development steps, validated several times during the project in an iterative manner and therefore, influence the development process directly.

IV. Results

The results of the project LICA contribute to the expansion and further development of innovative methods in data collection and participatory product development. New knowledge tools are going to be developed, which will be available for further research issues in the field of care and support after the end of the project. Now, during the project, data gaps and new (data) information regarding the various stakeholders in the care sector are identified and collected. On the one hand, this data refers to the situation and needs of people in care as well as professional caregivers in the informal or formal sector. On the other hand, this data refers to the communicative interaction between the care and the medical sector. Thus, it provides a valuable basis for the development of new projects in both, the nursing, and the care sector. Furthermore, the newly developed data sets (indicators) will enable previously unused possibilities for data exchange in the care sector and for the first time allow for a standardized data exchange between general medicine practices and pharmacies as well as therapists and other stakeholders without digital data discontinuity. The comprehensive involvement of different service providers in the development of the product offers a solid foundation for the adoption of the product and ensures a high-quality exchange of relevant information. The job satisfaction of those working in the health sector may increases due to reduced documentation effort, duplications, and losses of information – leaving more time for the care of the clients/patients themselves.

The inclusion of affected clients/patients and their relatives in the (ongoing) product development process, ensures the consideration of their needs and requirements. This increases the quality of care considerably in two regards: Firstly, the improvement of care and nursing on a somatic level is addressed and secondly the consistent participation of clients/patients and their social environment also increases the quality of care according to the WHO comprehensive concept of health:

Since physical as well as mental and social factors are equally included in the documentation process and clients/patients themselves can play a decisive role during the process, they experience themselves as individuals with the power to act, who can influence their environment and living conditions. Having the power to act represents a central resource for subjective well-being according to the salutogenic approach [1].

The interoperability between the project solution and other software systems, such as ELGA, general practitioners' software, pharmacies, mobile services, etc. is a novelty. The elaboration and application of indicators in the form of a care summary is an essential part of the content-related work and is only possible in a meaningful way by bringing different areas closer together. The increase in knowledge is possible due to the cooperation of mobile services, economy, scientific partners, and those affected.

V. Conclusion and Future Works

The LICA-project is innovative – for the first time the broad use of ICT and the strong networking of different organizations and people involved, will make it possible to assess, which practical benefits each of the intended functionalities will create for the mobile care services, caring relatives and how the user interfaces need to be designed to fit their needs. The user-centered design approach, which is applied in the whole research process enables this plan.

The profound literature research as well as the currently conducted focus group interviews show first significant results. These results provide a solid basis for the imminent research methods, which will take place in the upcoming four years.

Acknowledgment

This research is sponsored by the Austrian research promotion agency (FFG) under grant number 884194. Project partners are: Akademie für Altersforschung am Haus der Barmherzigkeit, Johanniter Österreich Ausbildung und Forschung gem. GmbH, Wiener Rotes Kreuz- Rettungs-, Krankentransport-, Pflege- und Betreuungsgesellschaft m.b.H., Volkshilfe Gesundheits- und Soziale Dienste GmbH (GSD GmbH), Volkshilfe Wien gemeinnützige Betriebs-GmbH, CareCenter Software GmbH, Loidl-Consulting & IT Services GmbH, Compugroup Medical CGM, Österreichische Apotheker-Verlagsgesellschaft m.b.H, Steszgal Informationstechnologie GmbH, Fachhochschule Technikum Wien, Universität Wien.

References

[1] A. Antonovsky: "Unraveling the mystery of health: how people manage stress and stay well". San Francisco: Jossey-Bass, 1987.

[2] M. Firgo and U. Famira-Mühlberger: "Expansion of inpatient care in the federal states. Quantitative and qualitative effects of the use of public funds compared to mobile care". WIFO. Vienna. http://www.wifo.ac.at/wwa/pubid/47447, 2014.

[3] E. Haslinger-Baumann, A. Lilgenau, C. Binder, and K. Gugenberger: "Quality characteristics in 24-hour care from the perspective of those affected. Results of a pilot survey study in Austria". In Pflege Die wissenschaftliche Pflegezeitschrift für Pflegeberufe, 2019.

[4] L. Jung-Joo: "Against Method: The Portability of Method in Human-Centered Design", Doctoral Dissertation, Aalto University, School of Arts, Design and Architecture, Department of Design, Helsinki, 2012.

[5] B. Mittermann: "Long-term care: Where new challenges meet an old system". Arbeit&Wirtschaft. https://www.arbeit-wirtschaft.at/langzeitpflege-herausforderungen/, 2019.

[6] M. Wright, M. Block, and H. V. Unger: "Levels of Participation in Health Promotion: A Model for Assessing Participation". In: Gesundheit Berlin (Hrsg.): Dokumentation 13. bundesweiter Kongress Armut und Gesundheit, Berlin, 2007.

[7] Care Network. https://www.carenetwork.org.uk/.

[8] CEF Digital Connectong Europe. https://ec.europa.eu/cefdigital/.

[9] My HealtheVet. https://www.myhealth.va.gov/mhv-portal-web/home.

[10] Siemens Healthcare Diagnostics GmbH. https://www.siemens-healthineers.com/at/digital-health-solutions/digital-solutions-overview/patient-engagement-solutions/e-health-solutions.

Effectiveness of a Biometric Patient Identification System

Eman Buhagiar, Clifford De Raffaele
Faculty of Science and Technology
Middlesex University
Malta
emails: buhagiareman@gmail.com, clifford.de.raffaele@gmail.com

Abstract— **Medical errors, such as patient misidentification, are the cause for around 2.6 million deaths per annum and $42 billion in cost to health organizations in low to middle-income countries. While wristbands are the most common method for identifying patients, they can be easily misplaced and may contain missing or inaccurate information as this study shows. This may result in wrong medications and surgeries and in some instances, even preventable deaths together with liabilities for health organizations. In this study, the current methods and process for identifying patients are investigated, accompanied by a comparison of existing patient identification solutions, as well as issues and concerns about health data protection and privacy. Following this, the effectiveness of biometric technology for patient identification is examined. The paper finally proposes and evaluates a proof of concept with promising results for minimizing patient identification errors.**

Keywords-medical errors; health organizations; biometrics; patient identification; face recognition.

I. Introduction

Patient misidentification is a recognized worldwide problem faced by medical organizations of different types and sizes [1], [2]. It is estimated that around 2.6 million people die each year, in just low to middle-income countries, due to medical errors [3], including errors of patient identification [4]. 9% of 7,600 (684) patient misidentification events captured in 181 different health organizations over the span of 32 months in the US led to patient harm, and in some cases, death [9]. Patient identification errors occur on different levels throughout the medical field. Various medical wards and units have been subject to such errors, including but not limited to, maternity wards, oncology centres, Intensive Care Units (ICU) and children's hospitals. In certain situations, such misidentification has led to severe consequences, one of which is the death of a patient [3][5]. According to the World Health Organization [3], between November 2003 and July 2005, the United Kingdom had 236 reported incidents related to missing wristbands or wristbands with incorrect information, the United States of America also had more than 100 similar cases reported from January 2000 to March 2003.

The National Patient Misidentification Report conducted by Ponemon Institute LLC in 2016 in the US [19] highlights the primary root causes of patient misidentification. The main three reasons include incorrect patient identification at the point of registration, time pressure when treating patients, and thirdly, lack of employee training and awareness. The report also outlines the health organization's financial impact, where the denial of claims costs the average healthcare organization $1.2 million a year. In a survey conducted by the same institute, seventy-six per cent (76%) of the respondents, who work in different types of organizations, such as large hospitals and small clinics, responded that biometrics at the patient registration point could reduce denied claims.

Patient misidentification may also lead to duplicate medical records that are time-consuming for organizations to manage and arrange [6]. An increase in insurance fraud for intentional misidentification may also be the cause of errors in patient identification. The National Health Care Anti-Fraud Association of the United States [7] estimates that the financial losses due to health care fraud are between 3-10% of the annual health care expenditure, which could lead to more than $300 billion a year. Moreover, according to the Medical Theft Alliance (MIFA), more than 2 million American citizens have been victims of medical identity theft, with cases rising each year [8].

This study analyses the effectiveness of using biometric technology for identifying patients by performing a literature review in Section II on the current problems caused by patient misidentification, followed by elucidating the process of identifying patients, current existing identification solutions, and the security and privacy issues and concerns regarding identifying patients. For the methodology in Section III, a list of system requirements is developed after distributing a questionnaire to a number of health professionals and analysing the responses. Once the requirements are documented, a system based on face recognition technology is proposed and designed in Section IV followed by its evaluation against a dataset in Section V. Finally, the results are analysed, and further improvements are suggested in Sections VI and VII.

II. Literature Review

Despite patient identification errors being preventable, many hospitals worldwide do not have patient identification systems implemented [13]. The first goal of The Joint Commission's National Patient Safety Goals (NPSG) for 2020 is to improve patient identification accuracy, both in hospitals and laboratories. Although the World Health Organization (WHO), the Emergency Care Research Institute (ECRI) [9], and other authors all promote the use of technology for reducing errors in patient identification [4][10][11]. It was found by the same ECRI that technology itself was the actual cause in 15% of patient misidentification errors. One of the potential barriers to mitigating or reducing patient identification errors is the costs associated with implementing such solutions [2].

A. *Identification Methods*

1) *Wristbands*

Full implementation of a barcode-based Electronic Positive Patient and Specimen Identification (EPSID) system can result in a significant reduction in mislabeled specimens over three (3) years [12]. However, other studies identified wristbands as one of the leading causes of patient misidentification [1], [4]. The main issues are missing or wrong information and patients having more than one wristband. Implementing a simple wristbands system for patient identification is considered a low-cost practice for health organizations [13]. Since the simple wristbands with handwritten information on them are still prone to human error, the use of barcodes [13] or RFID [14] can reduce or mitigate patient identification errors.

Efficacy of a barcode wristband system on the prevention of medical errors indicated that the system can reduce some medical errors by an estimated 12.22% to 57.4% in different hospitals [15], [16], and medication error rate by 56% and by 47% in neonatal intensive care units [17], [18]. A barcode wristband system can help such organizations in saving roughly $684,000 a year, from just denial of claims [19].

While wristbands are portable, relatively cheap, and generally easy to use, multiple problems can arise. One study concluded that 1 out of 84,000 barcode scans generated an incorrect patient identifier and as many as three (3) incorrect patient identifiers were outputted from a barcode [20]. Although this is a minimal number, these cases can still be fatal for a patient and costly for health organizations.

2) *Palm Vein Pattern Recognition*

Palm vein scanning is a widespread method of verifying and authenticating a user [21]. Given that each patient's palm vein pattern is unique and very stable over the person's lifetime, it makes this method the most commonly used successful technology for identifying people [22]. A palm vein scanner uses a near-infrared light wave to capture the user's vein pattern on the palm. In contrast with other recognition methods, palm veins have internal features making it almost impossible to reproduce with fake palms [23].

While taking into consideration the accuracy of the palm vein scanning method, it is worth noting that this method is more costly when compared to the barcode wristband alternative. This is due to its unique software in addition to the installation and the implementation of the palm scanners. This form of method is also considered to be more intrusive for a patient as it may raise palm image storage security concerns. However, when compared to the fingerprint or face recognition methods, the palm vein scanners are favorable within this regard. Another issue worth considering is the matter of hygiene as when comparing methods, a noncontact method would be ideal, examples of this include barcode scanning and face recognition.

3) *Ocular-based Identification*

Two types of ocular-based identification technologies used to identify a person uniquely are iris and retinal scanning. The retina is the thin tissue located at the back of the eyeball, containing cells sensitive to light. It is composed of a complex structure of capillaries that supply the retina with blood and therefore, every person's retina is unique. Similar to palm vein pattern recognition, a retinal scan would map a person's retina's unique patterns.

Iris recognition method would be ideal in a health organization environment as it does not require proximity to a camera for a successful scan and uses safer low-energy infrared lighting. Moreover, retina scan accuracy may be affected by certain diseases [24] and iris scanning proved to be the most secure patient identification method in UCSD's Moore Cancer Center when implemented [25].

4) *Face Recognition*

Face recognition can be described as determining the identity of an individual based on the person's facial features. The challenge of facial recognition in its simplest form involves comparing two face images and deciphering if they are of the same person [26]. A more significant challenge arises when faces exhibit changes in appearance due to make-up, facial hair, and accessories, such as jewellery.

One implementation of facial recognition with Microsoft Kinect v2 sensor for patient verification proved to be over 96% accurate [4]. However, each scan took around thirty (30) seconds to complete, a time frame which is unsatisfactory for a healthcare environment, but this can be classified as a limitation to the technology used, Microsoft Kinect v2, as other studies showed promising results in terms of performance, with time reduced to 100ms with the same level of accuracy [27][28].

B. *Security and Privacy*

The security and privacy areas in patient identification are habitually overlooked [29]. Privacy is also a significant concern for the patients themselves [31], and implementing a biometric system for improving patient identification accuracy is known to impose more privacy concerns for the patient [32].

1) *Health and Data Breaches*

According to a report issued by McAfee [33], a stolen health record would generally sell more than financial data on the black market. This is mainly because health data does not have that many established markets like financial data. Another study conducted by Infosec Institute (2015), shows that there was a 73% increase in cyberattacks between 2013 and 2014 targeted to healthcare organizations and that the average cost of a stolen health record amounted to $363 on the black market compared to $1 - $2 of the stolen credit card information. Health data breaches tripled in a year between 2017 and 2018 and there were over 15 million patient records breached in 2018 in the United States [34].

One of the most common causes of insider-related breaches is family member snooping [34], that is, healthcare workers spying on their family members. This cause amounted to around 67% of the breach cases, while the second most common type of breach was snooping on their co-workers, amounting to approximately 15% of the violations. Insiders, which are the healthcare workers, are also more likely to commit another breach after their first violation, as 51% of the offences are repeated.

2) *Privacy*

Storing and processing patients' personal and sensitive data calls for strict privacy protection measures to minimize patient privacy issues as much as possible. Biometrics privacy can be interlinked with personal privacy, given that our biometric information can uniquely identify us [35]. Various studies address different patient privacy concerns and implications [32][36]. In some cases where biometric technology is in place, patients refused to be subject to such technology due to privacy and confidentiality concerns [11]. Some biometric technologies proved to have a high acceptability rate, such as face recognition and voice recognition [4]. In contrast, others, such as iris and retina scanning [37], had a lower acceptability rate. Other studies however showed that biometric technologies are less or non-invasive than traditional methods of identification [4].

While there are no legislations covering the usage of biometric identification systems [38], and yet the right of privacy is considered a fundamental human right [39], safeguards must be set down for every step, from collection to retention of the data collected. Individuals must be given rights to access, correct and delete their data [35]. Furthermore, individuals should be assigned the ability to opt out, so biometric technologies should not be the only implementation for identification.

3) *Security*

Biometric technologies can help in identifying patients accurately and provide the right authorization and authentication or verification for accessing and amending medical records [26], [40]. The user asserts an identity for confirmation, and the biometric system confirms if the assertion is genuine. This process is generally used to prevent unauthorized access to a system or services. Verification can be explained formally using (1), where, given a claimed identity I and a query feature set x^A, the decision if (I, x^A) belongs to the 'genuine' or 'impostor' class needs to be taken. If x_I^E is the stored template that corresponds to the identity I, x^A is compared with x_I^E and a score s is matched, which measures the similarity between $x^A$and x^E, and η would be a predefined threshold.

$$(I, x^A) \in \begin{cases} geniune, \; if \; s \geq \eta, \\ impostor, if \; s < \eta, \end{cases} \tag{1}$$

However, biometric solutions can have their security flaws as well [41][42]. The biometric system's integrity is determined by its ability to guarantee non-repudiable authentication, that is, ensuring that a user who accesses a specific resource cannot later deny in using it. There are four major classes of security threats to biometric systems [6][26] and these are Denial of Service (DoS), Intrusion, Repudiation and Function Creep. Although it is much harder for an impostor to forge biometric traits than hacking traditional passwords, there are studies suggesting the use of multimodal biometric systems where multiple types of biometric features would be measured and compared, for example, fingerprints and face, for better accuracy [43][44].

The goal of this paper is to provide a proof of concept of the most favoured method of biometric patient identification determined through the methodology and evaluate its results against an already established dataset. Limitations and possible improvements are then suggested.

III. METHODOLOGY

A questionnaire was developed and distributed to professional healthcare participants using purposeful sampling, and its results are analysed. Consequently, requirements are determined, documented and validated using a House of Quality matrix and system designs are proposed.

A. Research Instrument

The questionnaire developed and conducted was sectioned into four (4) main sections:

- **Background Information** - gathering brief, non-personal information about the stakeholder, including their profession, roles, and practical experience.
- **The Problem** - capturing the stakeholder's awareness of patient misidentification and its consequences, globally and in the organization in which they practice in.
- **Their Process of Identifying a Patient in their Organization** – gathering information about the current process that health professionals use to identify patients. They were asked to explain the process briefly and what identifiers are used and at what point. They were also asked of awareness of any of the patient identification methods mentioned earlier in this study and which of them are used in their organization, if any. Finally, they were asked to provide feedback on their current patient identification method, and if they think that it can be improved and on what aspects, such as accuracy, security and cost.
- **The Solution(s)** - participants were asked which patient identification methods they would implement in their organization, how would they prioritize them and why. They were also invited to prioritize the characteristics and the concerns of a biometric patient identification system in terms of security, accuracy, and efficiency.

B. Participants

Participants chosen that successfully answered all the questions which were provided to them amounted to nine (9), and these were staff nurses (2), an Accident and Emergency (A&E) nurse, a doctor, a general practitioner (GP), a urology surgeon, physiotherapists (2), and a speech-language pathologist. All the participants work in local health organizations in Malta. The GP owns, manages, and works in a small private healthcare clinic. The nurses, the doctor, and the urology surgeon work in a national hospital, while the physiotherapists and the speech-language pathologist work in smaller private healthcare organizations.

C. Results Analysis

All the respondents think that their current patient identification system works moderately well (67%) or very well (33%). Hence, participants were also asked to identify the vital positive characteristics of their current patient identification system, and these included the cost, where 58% rated it as very well, followed by ease of use and efficiency (17%), and patient's comfort in using it, where 30% of the participants classified it as very well, as shown in Figure 1. Security was the least rated, with 33% rating it as just slightly well. 25% of the respondents classified security as an aspect that needs to be improved in their current system, along with accuracy. Moreover, 89% of the participants said that, currently, it takes less than 15 seconds to identify a patient, with 33% of them stating that it even takes less than 5 seconds. Therefore, essential requirements that needed to remain there are the system's cost, ease of use, patient's comfort in using it, and efficiency (processing speed). On the other hand, other aspects that require improvements are security and accuracy.

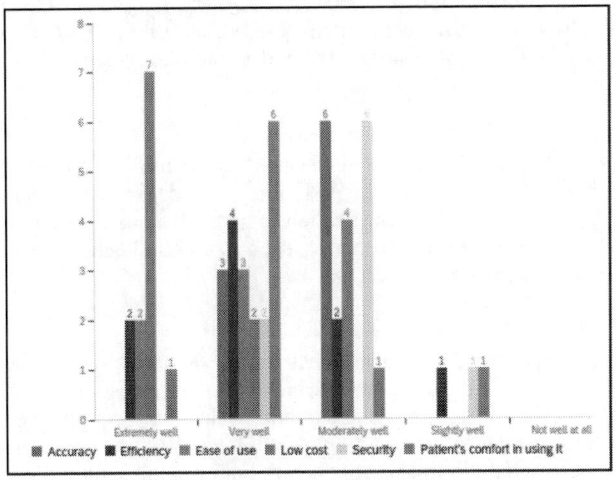

Figure 1. Question results for ranking aspects of system

D. Requirements

One of the most commonly used methods to achieve a standard view of the relationship between customer requirements and product design is Quality Function Deployment (QFD) [45]. QFD is a product development methodology that gives importance to the customer's opinions throughout the development process. QFD was used in this study to determine the list of important requirements for the proposed system. Customer importance ratings for system requirements were calculated based on the results obtained from the questionnaire distributed to health professionals in the order of Accuracy, Efficiency, Security, Ease of Use, Cost and Patient's Comfort. On the other hand, the requirements concluded from the previous section were listed on the other side of the matrix. The conditions that scored the highest importance ratings were found to be *Cost*, *Use of Secondary Identification Methods*, *Accuracy*, *Availability*, and *The Use of Alternative Non-Biometric Identification Methods*.

With the cost being the topmost essential requirement for the customer, any negative correlations related to this requirement should be addressed and ideally eliminated as soon as possible. Therefore, alternative non-biometric identification methods should be kept to a minimum and only used in cases where the patient refuses to use other biometric methods, for example. Patients will most probably opt for these alternative methods if they have trust concerns about the system, and hence the importance of *Transparency*. Each customer should be as transparent as possible to the patients about the biometric system, ensuring no physical harm will be done and securing their data safety while pointing out the benefits of such techniques for their own good. We must remember that using alternative non-biometric systems may negatively impact user training, the effort of operating the system, and identification accuracy.

IV. DESIGN

High-level and low-level designs of the system and integration with the possible current systems are proposed. Furthermore, designs of the proposed mobile application are also portrayed together with data and process flows. Taking Systems Theory into perspective, the proposed system would have biometric information as an input and after biometric processing and communication with the Patient Medical Record System (MRS) or Database, outputs the patient information.

For patient identification, the app user needs to be authorised and authenticated. The app should display multiple authentication choices, including but not limited to Face Recognition, Fingerprint Recognition, or a user account. Data collected at this stage is transferred securely to an internal API where it is processed. Through in-house or third-party APIs, if needed, roles and permissions are determined and set, and the user is then allowed to proceed and identify a patient on the app.

Provided the user is authenticated and authorized on the mobile application, the user can identify a patient, which has already been registered before, that is, the patient's biometric data required for identification has been securely stored on a database or service. After the patient has given consent and the biometric data is collected, such as a face photo, this data is sent through a secure and encrypted channel, such as HTTPS, to an internal API which communicates to third party APIs, such as Microsoft Facial Recognition, and handles the identification and the fetching of patient information and medical records if needed to be sent back to the app so that vital patient information can be displayed.

The users can be provided with different options to authenticate themselves (Fig. 3a). Viable options include face or fingerprint recognition or a user account.

a) Face Recognition - A

A similar process used to identify patients through face biometrics can be used here to identify and authorize a user.

b) Fingerprint - B

Most of the modern smartphones are equipped with an inbuilt fingerprint sensor. This may be used to authenticate the user. However, this may require more development effort to implement.

c) User account - C

A user account is also another option, although less preferred since it is more time-consuming to develop and maintain. The integration of Microsoft Azure AD will help in improving performance if Microsoft Face Recognition is used for patient identification.

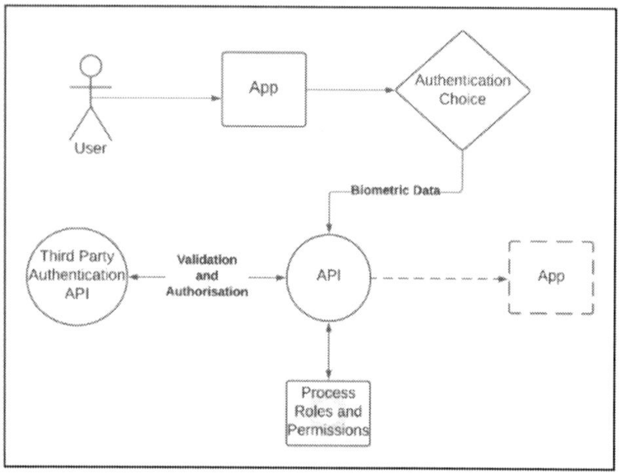

Figure 2. High-Level Authorisation Flow

Figure 2 shows the proposed flow of the mobile application used to identify patients by their face. Adopting two identifiers for identifying a patient, as suggested by the WHO [2], the first stage includes scanning the barcode or QR code printed on the wristband wearing the patient. The app should immediately display the camera preview after successful authentication, for the user to scan the barcode (Fig. 3b). The barcode should be recognised very quickly, and the Patient Identifier stored on the barcode or QR code is captured by the app. Once a barcode or QR code is successfully captured, the user should be prompted to capture the patient's biometric data. For this study, the method of face recognition using Microsoft Face Recognition is showcased. Therefore, the user is asked to take a photo of the patient's face, as straightforward as possible. The user should confirm the image taken for the identification process to initiate.

Upon identification completion, if succeeded, the user is prompted with a pop-up dialog asking to confirm the identification details, to ensure the identification and update biometrics or to scan again (Fig. 3c). Updating biometrics would send the last patient's face photo to Microsoft Face Recognition API and is added to the patient's list of faces for AI training.

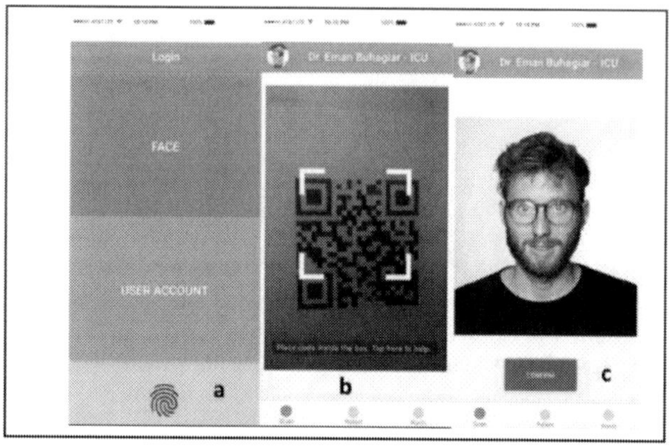

Figure 3. Proposed app system designs. a – Authentication, b – Barcode/QR code scanning, c – Identification confirmation

V. EVALUATION

The primary identification method of face recognition is implemented using Microsoft Cognitive Services and their Face API [46]. There are various reasons for opting for Microsoft Cognitive Services, and these reasons all cohere with the system requirements established earlier. Evaluation of the proposed system was divided into three stages:

A. Applicability

Microsoft Face API is a seamless, secure and an easy to integrate and operate API for face detection, emotion recognition, and identification. Microsoft Face API can be utilised in different scenarios, such as user authentication and counting people in a crowd.

1) Face Detection

Face Detection can detect up to 100 faces in an image along with different attributes such as age, position, smile, emotion, facial hair, makeup and occlusions, such as masks and bandanas when a photo or image URL is passed as a parameter. No images are stored, but only the landmarks are stored which cannot be used on their own to identify a person.

2) Face Recognition

Face Identification compares face landmarks previously stored on the API from adding faces to a person in a person group or large person group to an input image face landmarks. The API returns a confidence level (1-10) for the user to decide if the prediction is up to the user's expectations or not. The API also accepts a confidence threshold as a parameter to filter out results based on the user's preference for confidence. For example, in identifying patients, a confidence level below 80% (0.8) might not be acceptable, and therefore, a confidence threshold of 0.8 or greater should be passed.

3) Security

One of the most critical concerns addressed by health professionals in the questionnaire conducted for this study is the system's security. Microsoft ensures security by firstly not storing any actual face images on their servers, and secondly by encrypting any data stored using FIPS 140-2 compliant 256-bit AES encryption. FIPS 140-2 is a U.S. Government computer security standard used to approve cryptographic modules [47].

4) Cost

As for the cost of usage, the standard version allows for up to 10 transactions per second, with €0.506 per 1,000 transactions for 5 to 10 million transactions and €0.338 per 1,000 transactions for transactions amounting to more than 100 million. As for storage, €0.009 per 1,000 faces per month is charged. A transaction constitutes an API call, apart from the training calls where a transaction counts for every 1,000 images trained. The quota for the number of stored person groups is 1 million with up to 1 million persons per person group or face list. However, multiples of these groups can be created.

5) Limitations

Like all other face recognition methods, there are some limitations that may hinder the system's accuracy. Various face occlusions, such as masks and makeup, or face injuries and ageing, may prevent face recognition algorithms from detecting or identifying a face. While many face recognition technologies cater for occlusions measurement when detecting a face, face masks during the COVID-19 pandemic impacted face recognition algorithms' overall accuracy [48]. Such a situation requires the need for alternative identification methods, such as retina recognition, to be available in the system.

B. Accuracy Evaluation

A dataset of multiple faces was used to evaluate a person's identification accuracy through Microsoft Face API. The dataset was introduced in an another study [49] to provide more diversity than the existing publicly available datasets regarding lighting, age, and ethnicity. The dataset consists of 3755 faces, totalling to 276 participants in all. Each participant has at least eight (8) face photos, each from a different angle or different lighting (Fig. 4).

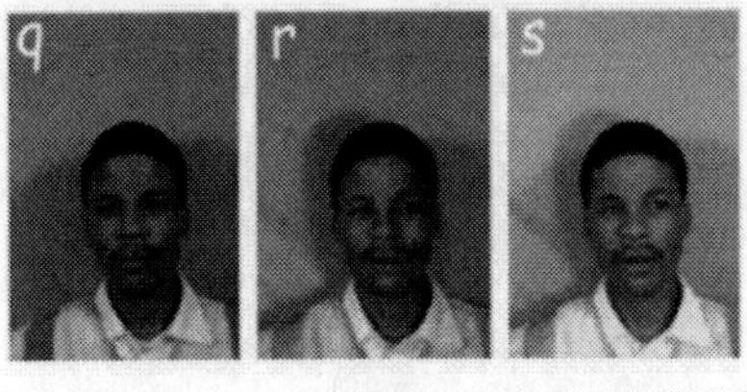

Figure 4. Face image under different lighting conditions

As a setup, a large person group was created, and for each of the subjects in the dataset, the subject was added to the large person group created and their face added to the person using their first image file. After all the subjects were added, the large person group was trained using the Train API endpoint.

Once all the subjects in the dataset were registered, the following identification process was conducted for each participant:

- The face was detected, and the *faceId* returned was stored.
- The face is identified, passing the *faceId* in the request body and the *largePersonGroupId* captured earlier when creating the group. If identified, a list of potential candidates should be returned, each with a *personId* and a confidence level.
- The person was identified and confirmed by getting the person in the large person group by the *personId* captured in the previous step. The person name and the file identification number were compared, and if these matched, identification was successful.

C. Performance Evaluation

The second stage of evaluation ensures that the second most crucial requirement established, efficiency is maintained throughout the identification process. For this, a simple mobile application was developed, simulating a patient's identification using two identifiers, a barcode and a face. Once this is done, the app prompts the user to take a photo of a face, and this is sent to Microsoft Face API upon confirmation for identification. Both the barcode key and the person identified from the API are compared, and if matched, a call to a database is made to fetch the records of the patient. The whole process was timed for efficiency evaluation.

VI. Results and Discussion

The results of the previously explained evaluation processes are analysed and discussed in detail. There are specific scenarios where the system performed very well, but there are others in which accuracy was challenged, and possible improvements are suggested for these cases.

A. Accuracy of Face Recognition

There were eight (8) types of datasets which were used to assess the accuracy of the proposed system. These will be referenced as *a, b, c, d, e* and *qa, rb, sb*. $a - e$ represent the different angles of the camera from which the photo was taken, while q, r and s represent the amount of light exposed to the face, with $s > r > q$.

The thresholds used for evaluating the accuracy of the system ranged from 0.97 to 0.92. Anything above the threshold of 0.97 resulted in less than 80% accuracy, which is not considered safe enough for such a critical system. On the other hand, any threshold below 0.92 always resulted in 100% accuracy in all scenarios tested.

1) Case 1 – Camera Angles

In the first case, the first set of faces *qa*, that is, photos taken from in front of the person (*a*) and with lighting set *q*, was added to the API and trained. After that, all the other photos from the remaining angles were tested for identification against different thresholds.

As seen in Table I, the accuracy results obtained by training just *qa* were always above 88% for 276 photos. It can be noted that angles *b* and *c* resulted in less accuracy than angles *d* and *e*. Therefore, side angles seem to be less accurate than front angles. Angles *c* and *d* provided the same accuracy results, while angle *d* proved to be the best angle for obtaining accurate results.

2) Case 2 – Lighting

In the second case, with trained set *qa*, all the other remaining photos with r and s lighting exposed to them were tested for identification with different thresholds.

In this case, accuracy suffered much more when different lighting was used on the person's face. As shown in Table II, the accuracy went down to 43.96% and 39.56% from 88.04 and above 99% from the previous case. This indicates that lighting has a significant effect on identifying a person from their face, and a less threshold of 0.92 compared to 0.94 had to be used for achieving 100% accuracy on the 91 photos tested. Lighting set *r* performed better than set *s* significantly. Significant changes were also noted when the threshold was changed each time by 0.01, with accuracy changes of more than 30% in some cases.

Since this case resulted in low accuracy results in some scenarios, set *ra* was added to the API and trained, and set *sb* was tested again. The same angle of the previously trained set was used (*a*) for consistency. This was done to note the difference in accuracy and the effectiveness of training. Table II below shows the accuracy results of set *rb* when tested, while sets *qa* and *ra* are trained already.

TABLE I. CASE 1 RESULTS

Thr. / Sc.	0.97	0.96	0.95	0.94
b	88.04	98.55	99.28	100.00
c	88.04	98.55	99.28	100.00
d	99.28	100.00	100.00	100.00
e	93.48	97.83	98.91	100.00

n = 276

Accuracy improved significantly for set *sb* when *ra* was added and then trained. With 0.97 as the threshold, accuracy improved by more than 24% and by more than 17% for the 0.96 threshold (Table III). This shows that dataset training provided by Microsoft Face API does improve identification accuracy.

TABLE II. CASE 2 RESULTS

Thr. / Sc.	0.97	0.96	0.95	0.94	0.93	0.92
rb	43.96	73.63	90.11	96.70	100.00	100.00
sb	39.56	64.84	82.42	93.41	98.54	100.00

n = 91 (photos 000 - 090)

B. Integration Efficiency Test

For this case, a simple mobile application was developed to showcase the use of the proposed system by the users. The app communicates with a custom developed API hosted on Microsoft Azure, which then communicates to Microsoft Face API and a database with records of patients, also hosted on Microsoft Azure. The process took between 5 to 7 seconds when timed in code, with full-bar Wi-Fi connectivity, to detect and identify the face through the API, and to get the patient's allergies and conditions list from a sample database. This result coheres with the efficiency requirement established earlier for identification to take not more than 15 seconds and ideally not more than 5 seconds.

TABLE III. CASE 3 RESULTS'

Thr. / Sc	0.97	0.96	0.95	0.94	0.93
sb	64.84	82.42	92.31	96.70	100.00

n = 91 (photos 000 - 090)

VII. CONCLUSION AND FUTURE WORK

We understand that patient misidentification is a known global problem in various health organizations, and it can lead to further complications to the patient and the organization itself. While biometric technology is applied in multiple sectors, such as authentication and security, payroll, and banking, there are fewer studies on the application of biometric technology for positive patient identification. This study conducted a questionnaire among different health professionals to determine the top concerns for implementing a biometric system in healthcare. These included security, accuracy, cost, and patient cooperation. While most of the participants were aware of some of the biometric methods for patient identification, none of them has ever made use of any of them but would consider in doing so, given a better accuracy rate and robust security. This questionnaire helped in determining and prioritising the proposed system requirements.

For the proof of concept, this study evaluated the implementation of face recognition biometric technology for identifying patients, as this was the most preferred biometric method chosen by the questionnaire participants. Microsoft Face API was used as the third-party provider for recognising faces, and the proposed system was evaluated against its accuracy and efficiency, among other requirements determined. While results were promising with over 80% accuracy in most cases, this technology seemed to lack in identifying faces with occlusions, such as different lighting. When more than one face photo from different angles and different lighting is registered and trained, accuracy was improved significantly.

As for future works, the system needs to be evaluated against a larger dataset with a larger variety of face occlusions, such as makeup, accessories, and face injuries to mimic real-case scenarios in health organizations. Further studies on the security aspects of the system are also important to be conducted to minimise the risks of malicious attacks on the system and gain more confidence from the system users.

REFERENCES

[1] S. W. Renner, "Wristband Errors in Small Hospitals," vol. 28, no. 3, 1997.

[2] S. Kelly, "The patient misidentification crisis," Health Manag. Technol., pp. 12–14, 2016.

[3] WHO, "WHO calls for urgent action to reduce patient harm in healthcare," Saudi Med. J., vol. 40, no. 10, pp. 1075–1076, 2019. [Online]. Available: https://www.who.int/news/item/13-09-2019-who-calls-for-urgent-action-to-reduce-patient-harm-in-healthcare (accessed June, 2021) .

[4] G. Lippi, L. Chiozza, C. Mattiuzzi, and M. Plebani, "Patient and Sample Identification. out of the Maze?," J. Med. Biochem., vol. 36, no. 2, pp. 107–112, 2017, doi: 10.1515/jomb-2017-0003.

[5] M. Jonas, S. Solangasenathirajan, and D. Hett, "Patient Identification, A Review of the Use of Biometrics in the ICU," 2014, pp. 679–688.

[6] T. H. Tase, E. R. S. Quadrado, and D. M. R. Tronchin, "Evaluation of the risk of misidentification of women in a public maternity hospital," Rev. Bras. Enferm., vol. 71, no. 1, pp. 120–125, 2018, doi: 10.1590/0034-7167-2017-0134.

[7] NHCAA, "The Challenge of Health Care Fraud - The NHCAA," 2015. https://www.nhcaa.org/resources/health-care-anti-fraud-resources/the-challenge-of-health-care-fraud.aspx (accessed Aug. 04, 2020).

[8] Ponemon Institute, "Fifth Annual Study on Medical Identity Theft," no. February, 2015, [Online]. Available: http://medidfraud.org/wp-content/uploads/2015/02/2014_Medical_ID_Theft_Study1.pdf (accessed June, 2021).

[9] ECRI Institute, "Patient identification errors," 2016. doi: 10.1016/j.enfcli.2011.07.006.

[10] M. A. Yahiro, "Biometrics Solutions in e-Health Security," Spine (Phila. Pa. 1976)., vol. 19, no. Supplement, pp. 2274S-2278S, 2015, doi: 10.1097/00007632-199410151-00004.

[11] N. Anne et al., "Feasibility and acceptability of an iris biometric system for unique patient identification in routine HIV services in Kenya," Int. J. Med. Inform., vol. 133, no. September 2019, p. 104006, 2020, doi: 10.1016/j.ijmedinf.2019.104006.

[12] P. D. Hain et al., "An intervention to decrease patient identification band errors in a Children's Hospital," Qual. Saf. Heal. Care, vol. 19, no. 3, pp. 244–247, 2010, doi: 10.1136/qshc.2008.030288.

[13] L. V. Hoffmeister and G. M. S. S. De Moura, "Use of identification wristbands among patients receiving inpatient treatment in a teaching hospital," Rev. Lat. Am. Enfermagem, vol. 23, no. 1, pp. 36–43, 2015, doi: 10.1590/0104-1169.0144.2522.

[14] A. Aguilar, W. Van Der Putten, and G. Maguire, "Positive Patient Identification using RFID and Wireless Networks,", 2006.

[15] M. Khammarnia, A. Kassani, and M. Eslahi, "The efficacy of patients' wristband bar-code on prevention of medical errors: A meta-analysis study," Appl. Clin. Inform., vol. 6, no. 4, pp. 716–727, 2015, doi: 10.4338/ACI-2015-06-R-0077.

[16] B. Wegerbauer, "Can barcoded wristbands improve patient safety ?," no. January, pp. 1–5, 2007.

[17] N. Dwibedi et al., "Effect of bar-code-assisted medication administration on nurses' activities in an intensive care unit: A time-motion study," Am. J. Heal. Pharm., vol. 68, no. 11, pp. 1026–1031, 2011, doi: 10.2146/ajhp100382.

[18] F. H. Morriss et al., "Effectiveness of a Barcode Medication Administration System in Reducing Preventable Adverse Drug Events in a Neonatal Intensive Care Unit: A Prospective Cohort Study," J. Pediatr., vol. 154, no. 3, 2009, pp. 363-368, doi: 10.1016/j.jpeds.2008.08.025.

[19] Ponemon Institute LLC, "2016 National Patient Misidentification Report Independently conducted by Ponemon Institute LLC Sponsored by Imprivata," no. December, 2016.

[20] M. L. Snyder, A. Carter, K. Jenkins, and C. R. Fantz, "Patient misidentifications caused by errors in standard bar code technology," Clin. Chem., vol. 56, no. 10, pp. 1554–1560, 2010, doi: 10.1373/clinchem.2010.150094.

[21] O. Akinsowon, B. Alese, and O. Adewale, "Infrared Capture of Palm-Vein Blood Vessel Patterns for Human Authentication," J. Internet Technol. Secur. Trans., vol. 3, no. 1, pp. 203–209, 2014, doi: 10.20533/jitst.2046.3723.2014.0027.

[22] Imprivata, "Improving Patient Care with Positive Patient Identification - White Papers - HealthITAnalytics," 2015.

[23] H. Setiawan and E. M. Yuniarno, "Biometric Recognition Based on Palm Vein Image Using Learning Vector Quantization," Proc. 2017 5th Int. Conf. Instrumentation, Commun. Inf. Technol. Biomed. Eng. ICICI-BME 2017, no. June, pp. 95–99, 2018, doi: 10.1109/ICICI-BME.2017.8537770.

[24] M. D. Abràmoff, M. K. Garvin, and M. Sonka, "Retinal Imaging and Image Analysis," IEEE Rev Biomed Eng, pp. 169–208, 2010, doi: 10.1109/RBME.2010.2084567.Retinal.

[25] B. N. Haile, "The Eyes Have It : Iris Biometrics Safely Identify UCSD Patients for Radiation Oncology Treatment," 2010.

[26] A. K. Jain, A. A. Ross, and K. Nandakumar, Introduction to Biometrics. Boston, MA: Springer US, 2011.

[27] R. Ranjan et al., "A Fast and Accurate System for Face Detection, Identification, and Verification," IEEE Trans. Biometrics, Behav. Identity Sci., vol. 1, no. 2, pp. 82–96, 2019, doi: 10.1109/tbiom.2019.2908436.

[28] S. Anila and N. Devarajan, "Simple and Fast Face Detection System Based on Edges," Researchgate.Net, no. May, 2010, [Online]. Available: https://www.researchgate.net/profile/Anila_Satish/publication/225292501_Simple_and_Fast_Face_Detection_System_Base

d_on_Edges/links/09e414fd75a23d2c1b000000.pdf (accessed June, 2021).

[29] G. C. Hembroff, X. Wang, and S. Muftic, "Providing an Additional Factor for Patient Identification Based on Digital Fingerprint," Aug. 2011.

[30] E. B. Heinlein, "Medical records security," Comput. Secur., vol. 15, no. 2, pp. 100–102, 1996, doi: 10.1016/0167-4048(96)89322-9.

[31] P. Houston, "Research Responses to Patient Privacy Concerns," 2017.

[32] C. L. Parks and K. L. Monson, "Automated Facial Recognition of Computed Tomography-Derived Facial Images: Patient Privacy Implications," J. Digit. Imaging, vol. 30, no. 2, pp. 204–214, 2017, doi: 10.1007/s10278-016-9932-7.

[33] C. Beek, C. McFarland, and R. Samani, "Health Warning Report," 2018. [Online]. Available: https://www.mcafee.com/enterprise/en-us/assets/reports/rp-health-warning.pdf (accessed June, 2021) .

[34] Protenus Inc., "PROTENUS 2019 BREACH BAROMETER," pp. 1–21, 2019, [Online]. Available: https://email.protenus.com/hubfs/Breach_Barometer/2018/2019 Breach Barometer Annual Report.pdf (accessed June, 2021).

[35] C. U. Ebelogu, O. Adelaiye, and F. Silas, "Privacy Concerns in Biometrics," no. July, 2019.

[36] D. Birnbaum, K. Gretsinger, M. G. Antonio, E. Loewen, and P. Lacroix, "Revisiting public health informatics: patient privacy concerns," Int. J. Heal. Gov., vol. 23, no. 2, pp. 149–159, 2018, doi: 10.1108/IJHG-11-2017-0058.

[37] N. Dahiya and C. Kant, "Biometrics security concerns," Proc. - 2012 2nd Int. Conf. Adv. Comput. Commun. Technol. ACCT 2012, pp. 297–302, 2012, doi: 10.1109/ACCT.2012.36.

[38] R. Gellman, "Privacy and Biometric ID Systems : An Approach Using Fair Information CGD Policy Paper 028 August 2013," Cent. Glob. Dev., no. August, 2013.

[39] A. Puri, "Privacy is a fundamental human right," pp. 1–3, 2013, [Online]. Available: https://www.financialexpress.com/archive/privacy-is-a-fundamental-human-right/1174213/ (accessed June, 2021).

[40] A. Dandashi and W. Karam, "Biometrics security and experiments on face recognition algorithms," 2012 IEEE Symp. Comput. Intell. Secur. Def. Appl. CISDA 2012, 2012, pp. 1-8, doi: 10.1109/CISDA.2012.6291532.

[41] S. O. Olatinwo, O. Shoewu, and O. O. Omitola, "Iris Recognition Technology : Implementation , Application , and Security," Pacific J. Sci. Technol., vol. 14, no. 2, pp. 228–233, 2013.

[42] J. C. Mazura, K. Juluru, J. J. Chen, T. A. Morgan, M. John, and E. L. Siegel, "Facial recognition software success rates for the identification of 3D surface reconstructed facial images: Implications for patient privacy and security," J. Digit. Imaging, vol. 25, no. 3, pp. 347–351, 2012, doi: 10.1007/s10278-011-9429-3.

[43] K. Sasidhar, V. L. Kakulapati, K. Ramakrishna, and K. KailasaRao, "Multimodal Biometric Systems - Study to Improve Accuracy and Performance," Int. J. Comput. Sci. Eng. Surv., vol. 1, no. 2, pp. 54–61, 2010, doi: 10.5121/ijcses.2010.1205.

[44] W. Dahea, "Multimodal biometric system : A review Multimodal biometric system : A review," no. November, pp. 25–31, 2018, doi: 10.13140/RG.2.2.34056.65287.

[45] R. J. Hauser, A. Griffin, L. R. Klein, M. G. Katz, and P. S. Gaskin, "Quality function deployment," 2010, doi: 10.4271/870272.

[46] Microsoft, "Facial Recognition | Microsoft Azure," 2020. https://azure.microsoft.com/en-us/services/cognitive-services/face/ (accessed May, 2021).

[47] D. L. Evans, P. J. Bond, and A. L. Bement, "Security Requirements for Cryptographic Modules," 1994, [Online]. Available: https://nvlpubs.nist.gov/nistpubs/FIPS/NIST.FIPS.140-3.pdf (accessed June, 2021).

[48] M. Ngan, P. Grother, and K. Hanaoka, "Ongoing Face Recognition Vendor Test (FRVT) - Part 6A: Face recognition accuracy with masks using pre-COVID-19 algorithms," 2020.

[49] S. Milborrow, J. Morkel, and F. Nicolls, "The MUCT Landmarked Face Database," Proc. Pattern Recognit. Assoc. South Africa, pp. 32–34, 2010.